VARIABLE CAPACITANCE DIODES

VARIABLE CAPACITANCE DIODES

The Operation and Characterization of Varactor, Charge Storage and PIN Diodes for RF and Microwave Applications

KENNETH E. MORTENSON, Ph.D.
RRC International, Inc.

Copyright © 1974

ARTECH HOUSE, INC.

Printed and bound in the United States of America.

All rights reserved. No part of this book may be reproduced or utilized in any form or by any means, electronic or mechanical, including photocopying, recording, or by any information storage and retrieval system, without permission in writing from the publisher.

Standard Book Number: 0-89006-015-0

Library of Congress Catalog Card Number: 74-189395

MODERN FRONTIERS IN APPLIED SCIENCE

*Noise Performance Factors
in Communications Systems
by W.W. Mumford and E.H. Scheibe*

*Laser Applications
by Dr. William V. Smith*

*Microwave Engineer's Handbook
by Ted Saad*

*Logarithmic Video Amplifiers
by R.S. Hughes*

*Parallel Coupled Lines and Directional Couplers
by Dr. Leo Young*

*Microwave Filters Using Parallel Coupled Lines
by Dr. Leo Young*

*Microwave Semiconductor Control Components
by Dr. Kenneth E. Mortenson & Jose M. Borrego*

*Phased-Array Antennas
by Dr. A.A. Oliner & G.H. Knittel*

*Data Modem Selection and Evaluation Guide
by V.V. Vilips*

*Infrared To Millimeter Wavelength Detectors
by Dr. Frank R. Arams*

*Adaptive Electronics
by Dr. Wolfgang W. Gaertner*

*Gallium Arsenide Microwave Bulk
and Transit-Time Devices
by Dr. Lester F. Eastman*

*Avalanche Transit-Time Devices
by Dr. George I. Haddad*

*Microwave Filters for Communication Systems
by C.M. Kudsia & M.V. O'Donovan*

*Radar Detection and Tracking Systems
by Dr. S.A. Hovanessian*

*Significant Phased Array Papers
by Dr. Robert C. Hansen*

*Stripline Circuit Design
by Harlan Howe*

Spectrum Analyzer Theory and Applications
by Morris Engelson & Fred Telewski

Bubble Domain Memory Devices
by Dr. Alan B. Smith

Advances in Computer Communications
by Dr. Wesley W. Chu

Electronic Information Processing
by Dr. William V. Smith

Radars
by David K. Barton
(5 Volumes)
    Vol. 1. Monopulse Radar
        2. Radar Equation
        3. Pulse Compression
        4. Radar Resolution and Multipath Effects
        5. Radar Clutter

Ferrite Control Components
by Lawrence R. Whicker
(2 Volumes)
    Vol. 1. Junction Circulators, YIG Filters
          and Limiters
        2. Ferrite Phasers and Ferrite
          MIC-Components

Microwave Integrated Circuits
by Dr. Jeffrey Frey

Microwave Diode Control Devices
by Robert V. Garver

Microwave Transistors
by Dr. E.D. Graham and Dr. Charles W. Gwyn

Active Filter Design
by Arthur B. Williams

The Entrepreneur's Handbook
by Joseph R. Mancuso
(2 Volumes)

                ARTECH HOUSE, INC.
                affiliated with Horizon House — Microwave, Inc
                610 Washington Street
                Dedham, Massachusetts 02026

to my family -
Dot, Dave and Tim

PREFACE

The variable capacitance diode or varactor has been with us since the inception of the crystal diode, most notably pressed into use with the early radar systems during World War II. Although its variable capacitance properties were little used in comparison to its nonlinear resistance properties, in actual fact some of the anomalous circuit behavior of those early systems were later explained in terms of parametric operation utilizing, in part, the variable capacitance nature of the diodes. With the advent of the P-N junction, the variable capacitance property of the diode became substantially more important as it was no longer so significantly shunted by junction resistance such that, at RF and higher frequencies, its Q-value was limited only by the series or body resistance. Because of this improvement the diode began to see some initial service as an RF and VHF voltage controlled tuning element in the '50's. With birth of the diode parametric amplifier, converter and harmonic generator in the late '50's, full exploitation of the varactor from RF to microwaves, and even millimeter waves, began and continues still. Augmenting the parametric circuit utilization of the variable capacitance diode came the entire microwave control activity in the '60's with substantial interest in switching, limiting and phase-shifting. With one extreme form of this diode, the PIN (possessing minimum capacitance change), large scale phased array antenna structures became possible and are still being constructed. Throughout this thirty-year period many technical articles and much commercial data on these types of diodes together with their applications have been written, but, to date, no text has been available which focuses solely on this diode, bringing together in a single volume all of the pertinent factors concerning its operation and characterization. It is to this purpose that this book is addressed.

In determining the content, arrangement and emphasis of the material contained in this text, three principal objectives were kept in mind with regard to the needs that such a volume would serve. First, to enable the circuit desinger to more creatively and effectively utilize the varactor and PIN by providing a full understanding of the device properties, operation and design. To be included here also would be sufficient device knowledge to anticipate resultant circuit operational problems and limitations and permit their rapid resolution. Second, to permit the device designer to gain a more direct correlation between the electrical properties as viewed by the circuit user and the device physics including the practical problems and limitations of providing specific electrical characteristics, power handling, mounting and packaging. Finally, to provide a bridge of understanding between the circuit and device design activities such that meaningful and effective communication be established for the optimization of the required components employing variable capacitance diodes.

To achieve these objectives the text was organized into six chapters which, if read sequentially, would take the reader through three stages of learning. The first, and largely descriptive, stage contained in Chapters I and II provides the reader with a definition of the device together with its more common variations or types and examples of its range of application. Augmenting this general information is a complete discussion of the physical operation of the device describing the nature and location of the variable capacitance and its associated losses for both reverse and forward biased states. The second, and more analytical, stage contained in Chapters III and IV provides the reader with a detailed treatment of the electrical characterization of the device element. Chapter III presents the small-signal properties, including topics on depletion layer capacitance, injection or diffusion capacitance, related losses and device figures of merit such as Q-value and cut-off frequency. Chapter IV presents the large-signal properties including topics on the varactor as a time-dependent element, as a charge storage element and, together with the PIN, as a switching element. The third and final stage, contained in Chapters V and VI, deals with the encumbrances and limitations on the device as imposed by packaging and power handling limitations including electrical, signal and thermal considerations.

As the content of this book was generated over a period of years, the author had the opportunity to discuss and work with many people active in microwave diode design, characterization and use. Although it would be impossible to acknowledge all such interactions, the author would be remiss if he did not extend his acknowledgement, appreciation and gratitude to many whose ideas, points of view and helpful discussions are undoubtedly contained herein. Such acknowledgements are made to the many former colleagues at Microwave Associates, Inc. including A. Uhlir, Jr., M. Hines, C. Howell, C. Genzabella, J. White, R. Tenenholtz, N. Brown and many others. Further acknowledgements are made to the members of the former I.R.E. Committee 28.4 — High Frequency Diodes, who generated the small-signal standards for varactors and with whom he served, including A. Bakanowski, N. Houlding, R. Harris, J. Kenney, J. Hilibrand and others. In addition, many other professional colleagues contributed through discussions, meetings and correspondence over the years including R. Garver, P. Penfield, Jr., R. Rafuse, J. Early, R. Ryder, W. Coffey and A. van der Ziel. Finally, special acknowledgement of the efforts of J. Borrego in suggesting and/or making some additions to this text as well as providing a general review and to Mrs. B. Mattimore for her steadfast efforts in typing this manuscript and the maintenance of its organization is gratefully made.

Kenneth E. Mortenson

TABLE OF CONTENTS

		Page
I.	**Introduction**	
	1.0 Definition	1
	1.1 Types	1
	1.2 Uses or Applications	3
II.	**Physical Operation**	
	2.0 General	9
	2.1 Reverse Biased Operation	10
	2.1.1 Depletion Layer Capacitance	10
	2.1.2 Sources of Losses	15
	2.2 Forward Biased Operation	18
	2.2.1 Injection Capacitance	18
	2.2.2 Injection Losses and Base Conductivity Modulation	25
III.	**Small-Signal Characteristics**	
	3.0 General	27
	3.1 Depletion Layer Capacitance	27
	3.1.1 Capacitive Properties	27
	Analysis of Depletion Layer and Resultant Capacitance	28
	Variation of Capacitance with Frequency	33
	Variation of Capacitance with Temperature	34
	Variation of Capacitance with Material Parameters	40
	3.1.2 Associated Losses	40
	Evaluation of Base Resistance	40
	Frequency Dependence of Base Resistance	44
	Temperature Dependence of Base Resistance	44
	3.1.3 Q-Value and Cut-Off Frequency	46
	Dependence of Cut-Off Frequency on Materials	48
	3.2 Injection or Diffusion Capacitance	50
	3.2.1 General Admittance Properties	52
	3.2.2 Capacitance and Conductance with Uniform Base	55
	3.2.3 Capacitance and Conductance with Graded Base	56

IV. Large-Signal Characteristics
- 4.0 General 61
- 4.1 Varactor as a Time-Dependent Element 62
  - 4.1.1 Time-Dependent Capacitance 62
  - 4.1.2 Time-Dependent Base Resistance 64
  - 4.1.3 Time-Dependent Based Figure of Merit . . . 69
    - Evaluation of Figure of Merit 70
  - 4.1.4 Limitations on Time-Dependent Evaluation. . 75
- 4.2 Varactor as a Charge Storage Element 77
  - 4.2.1 Charge Control Equations 77
  - 4.2.2 Circuit Model 79
- 4.3 Varactor or PIN as a Switching Element 81
  - 4.3.1 Device Design and its Two-States. 81
  - 4.3.2 RF Equivalent Circuits 84
  - 4.3.3 Other Characteristics 87

V. Packaging Considerations
- 5.0 General 91
- 5.1 Electrical Properties 91
  - 5.1.1 Parasitic Elements 91
  - 5.1.2 Package Equivalent Circuits 94
    - First Order of Approximation 94
    - A More Exact Lumped Model 96
    - The General Transformation
      Representation 96
  - 5.1.3 Package Limitations 99
  - 5.1.4 Special Packaged Structures 99
- 5.2 Thermal Properties 100
  - 5.2.1 Package Thermal Resistance 100
  - 5.2.2 Package Thermal Time Constant 101

VI. Power and Signal Handling Capabilities
- 6.0 General 103
- 6.1 Signal Considerations 103
  - 6.1.1 Small-Signal Limitations 103
  - 6.1.2 Large-Signal Limitations 109
- 6.2 Dissipative or Thermal Considerations 111
  - 6.2.1 Temperature Limitations 111
  - 6.2.2 Sources and Locations of Dissipation 113
  - 6.2.3 Thermal Models 114
  - 6.2.4 Calculation of Temperature Rise 117

VARIABLE CAPACITANCE DIODES

INTRODUCTION   1

1.0 Definition

The variable capacitance diode is a P-N junction which is generally used not for its rectifying properties but rather for its voltage dependent reactive properties provided primarily by the depletion layer of the junction but also by injected charge stored in the base region. By specifying the doping profile through the junction transition region, the dependence of the depletion layer width and thus the depletion layer capacitance on applied voltage can be controlled to suit the intended application. As the junction capacitance is the basic element of the diode to be used, all other facets of the diode are minimized. In particular, special emphasis is made to reduce all loss elements such as series or shunt resistances or fixed reactances which dilute the voltage dependent nature of the diode. To this end, not only are special junction and geometry characteristics employed in the fabrication of these diodes but also special packaging, particularly for use in the UHF and microwave regions, to minimize losses and the fixed, parasitic package reactances. Thus, the variable capacitance diode has become a special breed of diode tailored to perform particular circuit functions, generally at high frequencies, which make use of its reactive nature.

1.1 Types

There are many varieties of variable capacitance diodes with most of the differences in characteristics being controlled by differences in doping profile and base thickness relative to the depletion layer width. The two primary modifications in doping profile are concerned with the transition from the p-region to the n-region in the formation of the junction. If the doping transition is abrupt in comparison to the depletion layer width, then the unit is known as an *abrupt junction* diode which is usually obtained by an alloying or metallizing fabrication technique. If the transition is graded in doping throughout the entire excursion of the depletion layer from zero bias to breakdown, then the unit is known as a *graded junction* diode which is usually obtained by diffusion techniques. These junctions always possess capacitance values which depend inversely upon the applied (reverse) junction voltage to some exponent. The exponent for the abrupt junction is one-half; whereas, the exponent can range from one-third to nearly one-half for the graded junction as the grading is made to vary from a linear to a very steep exponential characteristic approaching an abrupt change.

There can exist all sorts of hybrid structures between these two basic forms of junctions or doping profiles, but perhaps only two are sufficiently significant to mention here. First, the most common form of hybrid is that which possesses some degree of grading such that at low reverse voltages the capacitance obeys the cube root of voltage law, but at higher voltages takes on the square root law as the depletion layer moves out into a homogeneously doped region. Such diodes usually result from post-diffusion of alloyed or metallized units or from shallow diffused units for which the substrate or base region is uniformly doped. The second and perhaps the most important form of hybrid is the so-called *hyper-abrupt junction*. This type of junction is formed by creating an abrupt junction on a substrate or base in which the doping decreases with distance from the junction transition region. In such a diode, the depletion layer continuously moves into a region of lighter and lighter doping as an increasing reverse voltage is applied. As a result, the junction capacitance decreases more rapidly than the inverse square root law with voltage exponents reaching as high as five for narrow voltage excursions.

Other factors in the fabrication of variable capacitance diodes which markedly affect their characteristics and therefore create additional types of units are the junction and base geometry and dimensions relative to the depletion layer width. Most variable capacitance diodes have plane junctions, at least with respect to the maximum value of the depletion layer width, and therefore behave in a one-dimensional manner. However, in some instances, spherical junctions can be created or the shape of the base region near the junction region so tailored that the depletion layer exhibits a three-dimensional behavior with resultant C-V laws which differ from those of the usual plane junctions. In addition to the junction geometry or shape, the relative base thickness of the diode is another factor which introduces new variations on the basic junction characteristics. For example, to minimize the losses or resistance associated with the variable capacitance diode, the semiconductor substrate or base region must be made as thin as possible, ideally to the extent that the depletion layer at the breakdown voltage extends completely through to the base ohmic junction or contact. This condition of depletion layer extension, which has been termed "punch-thru" by this author from transistor terminology,

Introduction

can be made to occur at reverse voltages considerably less than the breakdown if so desired. Such a punch-thru varactor or variable capacitance diode exhibits a nearly constant value of capacitance with voltage as the voltage is increased beyond the punch-thru value toward breakdown. Diodes of this type, which possess extremely small losses, are generally more readily fabricated using epitaxial wafers in which the epitaxial layer (base) is of higher resistivity or breakdown voltage as compared to the substrate material. An extreme case is the so-called PIN diode which is a punch-thru unit possessing a nearly intrinsic base region such that the depletion layer extends across the entire I-region with only a very small applied reverse bias, generally a very small fraction of the diode breakdown voltage. Such a unit exhibits most of its capacitance change, therefore, around zero bias with little or no change over most of its reverse bias region. Another structure frequently used is the one in which the junction is a Schottky barrier or metal semiconductor junction. Advantage of this structure is its ease of fabrication as compared to the diffused structure.

In summary, the variety of variable capacitance diodes as determined by their C-V laws is great but the most-used units by far are the plane, linearly graded or abrupt junction units with the least losses. These units have extremely narrow base regions and are often designed as punch-thru diodes, the extreme case of which is represented by the PIN diode.

1.2 Uses or Applications

The present-day variable capacitance diode finds wide application in many of today's circuit components particularly in the higher frequency range extending from VHF up into the microwave and millimeter wave region. The importance of these diodes to the circuit designer exists because of three fundamental characteristics. First, these diodes possess the active properties with useful Q values over a broad frequency range extending even up into the millimeter wave region. Second, these diodes exhibit virtually instantaneous response when kept in the reverse bias region and exhibit no measureable hysteresis phenomena. Third, such units are capable of handling relatively high RF power levels operating up to tens of kilowatts peak power and hundreds of watts average or CW power in the microwave region.

There are four basic modes of circuit operation employing the variable capacitance diode. The first and perhaps one of the oldest uses for the variable capacitance diode is the small signal capacitance which is controllable by a reverse bias voltage over an extended range of capacitance change. This mode of operation, as illustrated in Figure 1.1a, is basic to all tuning and continuous phase shifting operations where a continuous change in reactance is required and where the RF

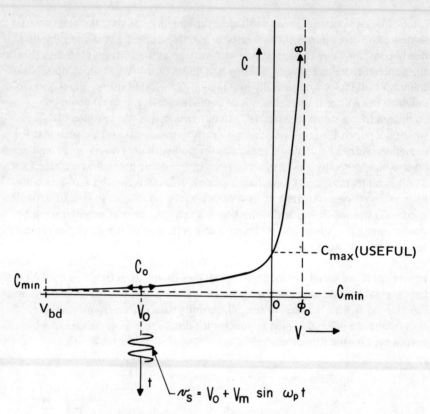

Fig. 1.1a Basic Operational Uses of the Variable Capacitance Diodes — Phase Shifting or Tuning Operation

signal level is sufficiently small that there is no reactive change with RF power level and little or no harmonic generation produced by the nonlinear characteristics of the diode.[1] In these applications, rapid tuning or continuous phase shifting can be accomplished by means of a low frequency control voltage at high speeds and with no hysteresis effects. The second mode of operation, which is perhaps an extreme case of the first, is that of switching or providing two impedance states, one approximating an open circuit condition and the other a short circuit condition as shown in Figure 1.1b.[2] The open circuited condition is approximated by a reverse biased diode which presents some capacitive reactance which can be parallel tuned to approximate an open circuit.

The short-circuited state is obtained by adequately forward-biasing the diode such that the junction impedance becomes very small with the resultant diode impedance being reduced to the series losses and parasitic elements. The third mode of operation is the pumped mode, wherein an RF signal with an appropriate bias is

Introduction

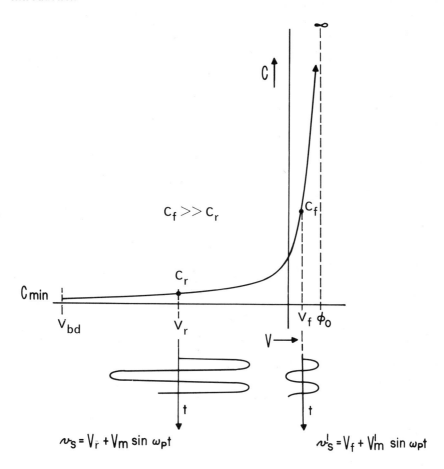

Fig. 1.1b Basic Operational Uses of the Variable Capacitance Diodes — Two-State Switching Operation

applied to the variable capacitance diode principally in the reverse bias region to achieve a time-dependent capacitance or a time-dependent storage of charge.[3,4,5] This type of operation illustrated in Figure 1.1c is employed in all parametric circuits such as amplifiers, converters or harmonic generators in which the primary source of RF power, namely the pump, is converted into a useful output signal at a new frequency by virtue of the conversion processes obtainable from the nearly lossless time-dependent reactive element. The fourth and final mode of operation is that of power limiting[6] which results from the nonlinear characteristic of the variable capacitance diode such that, as the signal level is increased from a small value to a large one in terms of the applied junction voltage, the impedance level presented by the diode diminishes from some finite capacitive value to a nearly short-circuited state. This mode of operation as shown in Figure 1.1d

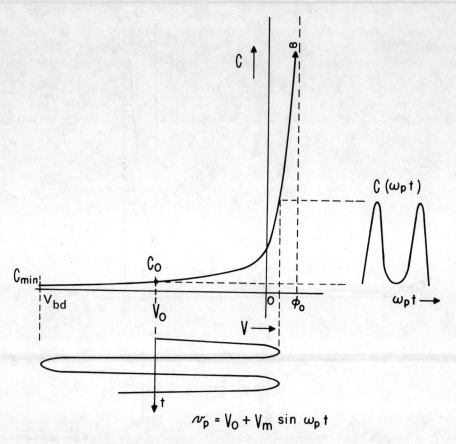

Fig. 1.1c Basic Operational Uses of the Variable Capacitance Diodes –
Pumped Parametric Operation

is entirely passive in that no bias is applied and the change in impedance state takes place by virtue of the change in RF power level. By virtue of the zero bias operation used in this application, both reverse and forward bias region characteristics of the diode are used. In summary, the basic modes of operation of the variable capacitance diode are continuous and step bias operations with small-signal RF voltages, RF pumped operations, and RF power level impedance change operations.

REFERENCES

1. M.H. Norwood and E. Shatz, "Voltage Variable Capacitor Tuning – A Review," *Proc. IEEE*, 56, 788-797 (May 1968).

Introduction

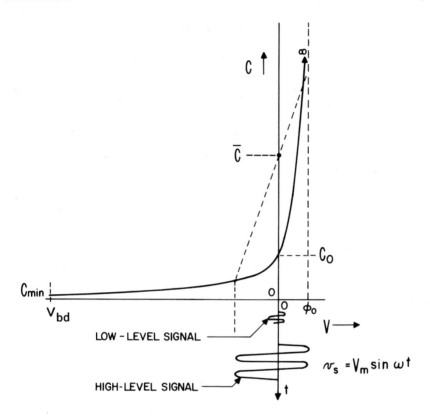

Fig. 1.1d Basic Operational Uses of the Variable Capacitance Diodes — Limiter Operation

2. R.V. Garver, "Theory of TEM Diode Switching," *IRE Trans. on Microwave Theory and Techniques,* MTT-9, 224-238 (1961).

3. P. Penfield and R.P. Rafuse, "Varactor Applications," M.I.T. Press, Cambridge, Mass. (1962).

4. M. Uenohara, "Cooled Varactor Parametric Amplifiers," in *Advances in Microwaves,* Vol. 2, ed. L. Young, Academic Press, Inc., New York, N.Y. (1967).

5. J.O. Scanlan, "Analysis of Varactor Harmonic Generators," *Advances in Microwaves,* Vol. 2, ed. L. Young, Academic Press, Inc., New York, N.Y. (1967). N.Y. (1967).

6. N.J. Brown, "Design Concepts for High-Power PIN Diode Limiting," *IEEE Trans. on Microwave Theory and Techniques,* MTT-15, 732-742 (1967).

BIBLIOGRAPHY

Blackwell, L.A. and Kotzebue, K.L., "Semiconductor-Diode Parametric Amplifiers," Prentice-Hall, Inc., Englewood Cliffs, N.J. (1961).

Chang, K.K.N., "Parametric and Tunnel Diodes," Prentice-Hall, Inc., Englewood Cliffs, N.J. (1964).

Levine, S.N. and Kurzrok, R.R., "Selected Papers on Semiconductor Microwave Electronics," Dover Publications, Inc., N.Y., N.Y. (1964).

Mortenson K.E. and Borrego, J.M., "Design, Performance and Applications of Microwave Semiconductor Control Components," Artech House, Inc., Dedham, Mass. (1972).

Nergaard, L.S. and Glicksman, M., "Microwave Solid-State Engineering," D. Van Nostrand Company, Inc., Princeton, N.J. (1964).

Shurmer, H.V., "Microwave Semiconductor Devices," Wiley-Interscience, N.Y., N.Y. (1971).

Uhlir, A. Jr., "The Potential of Semiconductor Diodes in High Frequency Communications," *Proc. of the IRE*, 46, 1099-1115 (June 1958).

Watson, A.A., "Microwave Semiconductor Devices and their Circuit Applications," McGraw-Hill, Inc., N.Y., N.Y. (1969).

PHYSICAL OPERATION   2

2.0 General

The purpose of this chapter is to discuss and explain the physical operation of variable capacitance diodes before getting into their detailed characteristics and analysis. A qualitative explanation of the operation of the P-N junction as a variable capacitance element will be given to answer such questions as where the capacitance of the junction originates, what charge is stored and where, what are the limits on this junction capacitance and how does it depend on voltage and frequency, and finally what losses are associated with this reactive element and where they exist in the diode. It will be assumed in this discussion that the general V-I characteristic of the P-N junction is known, that is, what the characteristic is, what is meant by the forward and reverse directions, and the phenomena of breakdown which restricts its voltage range (see Bibliography at end of chapter for a basic treatment of P-N junction characteristics).

The discussion presented in this chapter is divided into two principal parts to separate two basic sources of junction capacitance; namely, that produced by the placement of majority carriers in the transition region of the junction from that obtained by the injection of minority carriers into the base region of the diode. Thus, the discussion will first concern itself with the reverse biased operation which is limited to movement of majority carriers which results in the so-called depletion layer capacitance and its associated losses. The second portion of the discussion will be centered on forward biased operation where the action of injected minority carriers predominates and results in producing some form of injection capacitance as well as modifying the losses which are normally associated with the total junction capacitance.

2.1 Reverse Biased Operation

2.1.1 Depletion Layer Capacitance

In the formation of a P-N junction two regions of semiconductor material which possess opposite types of conductivity are brought together; that is, one possesses holes as its majority carrier as a result of ionizing acceptors impurities, and the other possesses electrons as its majority carrier as a result of ionizing donor impurities. As a result of this difference in conductivity type in the same semiconductor material a difference in Fermi level or work function exists such that in the fabrication of a P-N junction a contact potential difference is created between the two regions of the semiconductor body. This contact potential difference is in effect nullified by the flow of electrons from the N to the P region and the flow of holes from the P to the N region until the Fermi levels for the two regions align themselves. In this process the mobile carriers, that is, electrons and holes, are swept out of the volume existing in the metallographic transition from one doped region to the other, resulting in unneutralized ion centers held fast in the lattice structure of the semiconductor on either side of the interface between P and N regions. These unneutralized acceptor (-) and donor (+) ions form a dipole layer at the interface of the junction and thus represent a storage of charge as a result of the contact potential difference, ϕ_0. Shown in Figure 2.1 are two of the most common doping profiles for P-N junctions with their resultant depletion and dipole layers. For both of these cases, i.e., the abrupt and linearly graded junctions shown in Figure 2.1a and 2.1b respectively, x_0 represents the interface between the P and N regions forming the junction, and the shaded regions closest to this interface represent the volume which is depleted of mobile carriers as a result of the contact potential difference, ϕ_0. The dipole layer resulting from the depleted regions, as has been described, is shown in the adjoining plot of charge density versus distance perpendicular to the junction interface. The spatial distribution of charge density making up the dipole layer is determined primarily from the doping profile of the junction as it is fabricated. Thus for the abrupt junction case the dipole layer is quite asymmetrical about the interface plane; i.e., most of the charge on the heavily doped side of the junction exists very close to the interface surface; however, in the lighter doped region on the opposite side of the junction the charge is distributed over a considerable distance into what normally constitutes the base region of the diode, since the total charge on either side of the junction must be equal. For the linearly graded junction case, however, the spatial distribution of charge is symmetrical about the junction center and extends equally into both regions.

Physical Operation

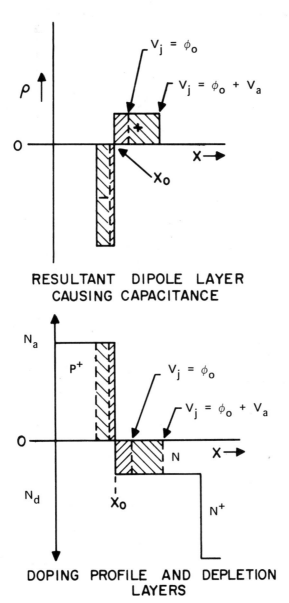

Fig. 2.1a Source of Junction Dipole Layer and Resultant Voltage Dependent Capacitance — Abrupt Junction Case (Idealized)

VARIABLE CAPACITANCE DIODES

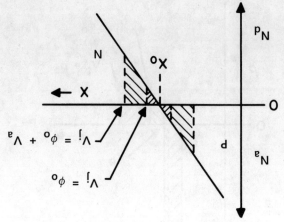

DOPING PROFILE AND DEPLETION LAYER

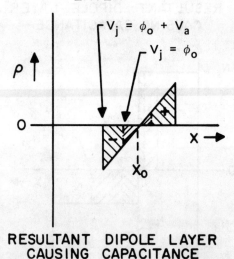

RESULTANT DIPOLE LAYER CAUSING CAPACITANCE

Fig. 2.1b Source of Junction Dipole Layer and Resultant Voltage Dependent Capacitance — Linearly Graded Junction Case (Idealized)

Physical Operation

So far the discussed junction depletion and dipole layers have resulted solely from the contact potential difference existing between the P and N regions. If an external voltage is now applied to the terminals of the diode, this voltage will be either added to or subtracted from the contact potential difference and will give rise to a new net junction voltage. If a reverse external voltage is applied to the diode the Fermi levels in the two regions are further separated, enhancing the action already brought about by the contact potential difference and thereby further depleting the region in the immediate vicinity of the junction and thus increasing the depletion layer width and the moment of the dipole layer. The change brought about by the application of a reverse potential is illustrated for both junction cases in Figure 2.1. The changes in the depletion and dipole layers resulting from the application of the reverse voltage are illustrated by the additional shaded regions. If the externally applied voltage is in the forward direction, the Fermi levels of the two regions are brought into closer alignment thus reducing the depletion layer width and the resulting dipole moment. Indeed, if the applied voltage could be made equal to the contact potential difference, the depletion layer width and dipole moment would be reduced to zero and the contact would in effect become ohmic. Such a condition of forward bias can never exist due to the infinite current which would be drawn through some finite ohmic losses in series with the junction resulting in the requirement of an infinite forward voltage. In general, variable capacitance diodes are not used at extreme forward bias.

Having described the formation of the depletion layer and resulting dipole moment and the variations in these layers due to the application of an externally applied voltage, it is next important to recognize the existence of an effective capacitance resulting from the charge stored within the dipole layer. In fact the incremental capacitance per unit area of such a dipole layer is simply expressed in the form of a parallel plate capacitor; i.e., equal to the dielectric constant existing between the separated charge divided by the charge spacing or, in this case, the depletion layer width. A schematic of the physical diode as it would appear for an abrupt junction is depicted in Figure 2.2 illustrating the depletion layer and its electrical equivalent, the voltage-dependent capacitance, $C_j(V)$. The depletion layer increases, in this instance, primarily at the expense of the moderately doped N type base region as a reverse voltage is applied. Thus the observed junction capacitance has been decreased from its zero bias value as a reverse voltage is assumed to be applied and the depletion layer has been widened, separating the charge of the dipole moment further. It is important to note here that one of the primary differences between the various junction types, as specified by the doping profile, is the dependence of the depletion layer width on the applied reverse voltage. For example, in the abrupt junction case, as an external reverse voltage is applied, for each additional increment of voltage, the same magnitude of mobile charge density must be swept out; whereas, in the linearly graded junction case, the charge

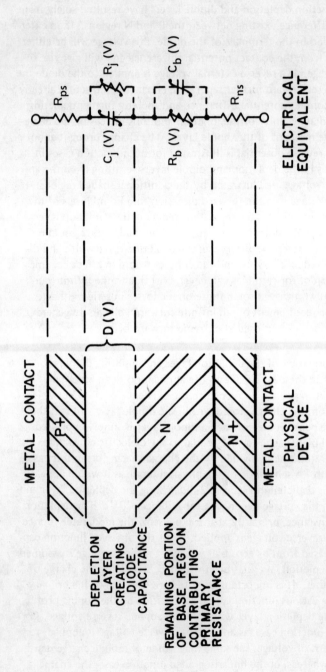

Fig. 2.2 Physical and Electrical Schematic of Reverse Biased Variable Capacitance Diode

Physical Operation

density to be swept out increases with every additional increment of applied voltage. This difference in the mobile charge density which is swept out as the depletion layer is widened yields different voltage dependences for the depletion layer width and thus different depletion layer capacitance-voltage relationships depending upon the built-in profile of the junction formed.

As the depletion layer capacitance so described is dependent only upon the motion of majority carriers on either side of the junction as the depletion layer is widened or narrowed with the applied voltage, the response of this capacitance to the applied junction voltage is nearly instantaneous. Thus, this voltage-dependent capacitance provided by a semiconductor junction is extremely fast in response and possesses no hysteresis effects with respect to voltage. Indeed, the response of this form of junction capacitance which involves no minority carriers is limited only by the time lag between the application of the external voltage to the diode and that voltage being established across the junction. This time lag is simply determined by the RC time constant, which is equal to the change in the depletion layer capacitance times the equivalent resistance in series with the junction. Thus the typical speed of response for such a variable capacitance diode is typically in the range of 10^{-10} to 10^{-13} seconds, corresponding to frequencies in the millimeter and submillimeter wave range. As a result, the depletion layer capacitance exhibited by these diodes is generally invariant with frequency over the entire frequency range up to and including microwave frequencies. Also, due to the fact that only majority carriers are involved, this form of capacitance is relatively insensitive to temperature over a wide range of temperature values above and below room temperature. In fact, the principal effect of temperature on the depletion layer and the resulting capacitance is in changing the value of the contact potential difference, ϕ_0, as the Fermi levels are shifted in each of the semiconductor regions making up the junction. Thus for reversed bias operation the junction capacitance is most sensitive to temperature changes near zero bias and becomes relatively invariant with temperature at high reverse biases; i.e., when the applied potential is very much greater than the contact potential difference.

2.1.2 Sources of Losses

In conjunction with the depletion layer capacitance created from a semiconductor junction in reverse bias are the inevitable losses which prevent the diode from being used as a completely reactive element. The sources of these losses are several, but by far the most predominant loss is in the base region of this type diode. Referring again to Figure 2.2, it is shown how, for the abrupt junction case illustrated, the remaining portion of the base (N type region) results in the series base resistance R_b. This base resistance is the largest contributor to the losses of this variable capacitance element principally because the doping or conductivity of this region must be made small enough such that the depletion layer can expand into this region to provide the voltage-dependent nature of the depletion layer capacitance. If the base region of this diode structure (exclusive of the depletion layer) were highly doped the capacitance would remain essentially constant with further increases in reverse voltage and the unit would become essentially a fixed capacitor. Thus the undepleted base region is a basic necessity if one is to obtain a variable capacitance diode and can only be minimized by controlling the thickness of the base region with respect to the depletion layer width. If indeed the base width is made sufficiently thin such that the depletion layer extends completely through this region to the ohmic contact at the breakdown voltage, then the resistance contributed by this region will vary from some finite value at zero bias to, ideally, a zero value at breakdown, resulting in a variable capacitance junction with the highest possible quality factor consistent with the full range of capacitance change obtainable for the breakdown voltage of the junction. In all but the highest quality variable capacitance diodes, which do not exhibit depletion layer punch-thru, i.e., possess a full range of capacitance change, the base region is thick enough such that little change in the base resistance is obtained as the depletion layer width is altered by biasing from zero to breakdown voltage. Thus, although the base resistance can be voltage dependent, it typically is not. Similar comments regarding the base resistance pertain to other junction types even though the detail dependence on doping profile and base thickness may be different.

Having discussed the voltage dependence of the base resistance, it is also important to note the dependence of this quantity upon frequency. Although the base resistance of a variable capacitance diode is generally invariant with frequency, at least up to the microwave frequencies, there are two effects which can appreciably alter the effective base resistance depending upon the frequency of interest, geometry and resistivity of the base region. For example, for relatively low breakdown voltage diodes, i.e., of the order of a few tens of volts, the base region is fabricated from relatively low resistivity material and is of such dimensions that skin effects can prevail, forcing the current to flow in a restricted region of the base thus increasing the effective series base resistance. This phenomenon is generally not important, however, unless the operating frequency is in the millimeter wave region or higher for typical variable capacitance diodes. The second effect, which generally pertains to the higher breakdown voltage diodes or units which possess a relatively high base resistivity, is that associated with the dielectric relaxation phenomenon. This phenomenon, which is basically concerned with the bulk impedance of the base region material as determined by *both* the conductive as well as the displacement currents, is schematically represented in Figure 2.2 as a voltage dependent, shunt capacitance appearing across the base resistance, namely $C_b(V)$. This capacitance is simply the capacitance of the base region as determined by the dielectric constant of the semiconductor material employed with no mobile carriers present. The magnitude of this capacitance is therefore obviously dependent upon the

Physical Operation

applied voltage in the same manner as the base resistance in that it is controlled by the net base thickness as the depletion layer width is changed. The effect of this phenomenon, expressed in terms of this shunting capacitance, is to reduce the series resistance of the base region as the frequency of operation is increased. However, this decrease is gained at the expense of the variation in junction capacitance with voltage, since at sufficiently high frequencies, the impedance between the ohmic contacts (in this case the P^+ and N^+ regions) becomes essentially capacitive and depletion width has no significance. For the most typical varactors or variable capacitance diodes possessing no punch-thru characteristics and breakdown voltages of 100 V or less, this effect is also not appreciable until the operating frequency occurs in the millimeter wave range or higher. However, for variable capacitance diodes which are fabricated from very high resistivity materials, this effect can exist even down into the microwave and UHF region. Indeed, for the extreme case of a PIN structure with an intrinsic base region, the dielectric relaxation phenomena becomes important in the tens of megahertz region. In summary, the undepleted base region of the variable capacitance diode contributes the largest portion of the losses as represented by a series base resistance which, as has been pointed out, is typically invariant with respect to applied voltage and frequency, but in some instances can be markedly dependent upon these parameters.

The next most important sources of losses are the series resistances indicated in Figure 2.2 as R_{ps} and R_{ns}. R_{ns} includes the body resistance of the N^+ or degenerate semiconductor region and the metal contact between the base and the external connection to the diode. Both these sources of series resistance will generally exhibit skin effects: the metal contact from very low RF frequencies on the order of 1 MHz on up and the degenerate regions on the order of 1 - 2 GHz on up for the dimensions typically encountered in these portions of the diode. Also included in R_{ns} is any possible contact resistance or voltage drop occurring between the N^+ and N region, since this contact to the base is not truly ohmic and can present a barrier drop of several hundredths of a volt or more depending upon the extent of the difference in doping of the two regions involved. In a similar manner, R_{ps} involves the resistance contributions of the other degenerate semiconductor region, the P^+ region, and the metal contact from this portion of the semiconductor element to the external diode connection. The sum of these body and metal conductor resistances can readily be several tenths of an ohm or higher at microwave frequencies and therefore represents a significant loss factor for very high Q variable capacitance diodes for which the base resistance is in the order of an ohm.

The final source of loss for the diode is that contributed by bulk and surface leakage through and around the depletion layer; this represents a conductive loss as represented by $R_j(V)$. This leakage resistance is generally negligible for variable capacitance diodes from the UHF frequencies on up because the leakage currents are relatively small and they are significantly shunted by the depletion layer capacitance at these frequencies. At lower frequencies, however, this resistance does become significant and indeed becomes the limiting effect controlling the Q of the variable capacitance diode as the frequency is lowered.

Thus, the losses of this type of diode result from contributions in many portions of the structure with the predominant contribution resulting from the undepleted base region of the semiconductor element. Except for the reverse biased leakage resistance, all loss elements appear in series with the junction capacitance and therefore can be conveniently lumped into one total, effective, series resistance. This series resistance can also be made to include the series equivalent of the shunt leakage resistance if this is not negligible as well as any shunt resistances contributed by the packaging of the semiconductor element. In this manner the variable capacitance diode, biased in the reverse direction, can be simply represented by the variable junction capacitance in series with an effective resistance which defines the Q or quality factor of the unit.

2.2 Forward Biased Operation

2.2.1 Injection Capacitance

In addition to the charge stored, which results in the depletion layer capacitance, injected minority carriers resulting from a forward applied voltage provide another source of charge storage and resultant capacitance. This capacitance, which adds to the depletion layer capacitance under forward bias conditions to yield the total junction capacitance, results not from a dipole layer associated with the junction but rather from the storage of minority carriers in the base region of the diode. The effects of such an injected charge is depicted schematically in Figure 2.3 which shows both the physical device and its electrical equivalent for the forward biased operating condition. In this figure, the junction is now represented by a capacitance and resistance in parallel; both of which depend on the injected current in series with the base resistance, which is also a function of the current. The equivalent electrical parameters are more directly dependent on the current than the voltage for forward biased operation; these conditions are just the opposite of those prevailing for reverse biased conditions. It is useful to consider the injection capacitance under two different applied voltage operating conditions; first, the case when a small ac signal is superimposed on a forward dc bias voltage, and second, the large ac signal case when typically no dc bias or possibly a reverse bias is employed. Each of these operating conditions will now be discussed to illustrate those facets of forward biased operation which are pertinent to a variable capacitance diode.

Physical Operation

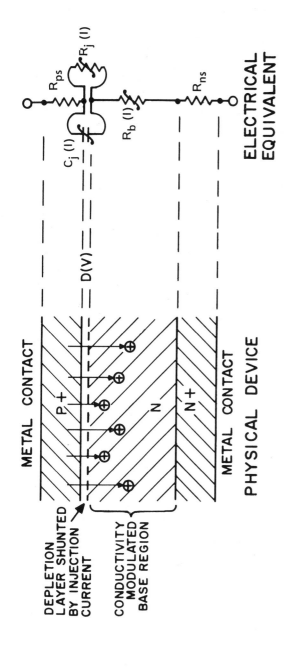

Fig. 2.3 Physical and Electrical Schematic of Forward Biased Variable Capacitance Diode Indicating Effects of Injected Carriers

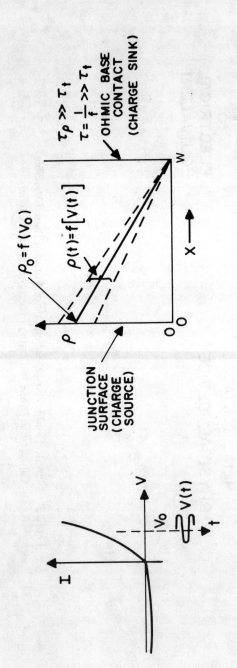

Fig. 2.4a Injected Minority Carrier Base Charge Density Distributions — Resultant Small Signal, Forward Biased Base Distribution for the Applied Voltage Condition Shown

Physical Operation

The first and perhaps better understood form of injection capacitance, better known as diffusion capacitance, occurs when there are small perturbations in charge stored in the base region of the diode, as depicted in Figure 2.4a. In this situation, when a forward bias is applied to the junction, an equilibrium charge density distribution is created with the junction becoming the charge source and typically the ohmic contact the charge sink such that a linear charge gradient exists across the base region as shown in the figure. This type of minority carrier base charge distribution will exist provided the doping in the base region is uniform and the lifetime of the minority carriers is significantly longer than the transit time from junction to ohmic contact. Under these conditions of forward bias, the charge distribution under equilibrium conditions represents stored charge which if perturbed by a small ac signal superimposed on the dc as shown in the figure will yield a change in stored charge for the given change in applied voltage which is, by definition, a source of capacitance. This portion of the total junction capacitance becomes predominant compared to the depletion layer capacitance at an applied forward voltage where the current density typically ranges from 1 – 10 A/sq. cm. and greater. This additional small signal capacitance which is created by the diffusion of carriers into the base region (hence the name diffusion capacitance) also has losses associated with it as is indicated in Figure 2.3 in the form of $R_j(I)$. This shunt junction loss is simply the conduction current associated with the injection or diffusion process and is obviously extremely important in determining the effective Q of the junction capacitance under forward biased operating conditions.

To further understand the operation of the forward biased variable capacitance diode, it is important to explore the relative importance of the real and imaginary portions of the junction admittance as produced by the diffusion or injection currents. Specifically, at low frequencies when the transit time of minority carriers across the base region (the lifetime τ_ρ is assumed to be much greater than the transit time τ_t) is less than approximately one-tenth of the period of the ac signal, the conductive portion of the diffusion current predominates and effectively represents the junction impedance, except for the depletion layer capacitance, which is generally also negligible at these frequencies. Not only does this junction resistance, R_j, dominate in this frequency range but its value is also independent of frequency. To appreciate what frequency range is typically involved under these operating conditions, one must first know that the value of transit time for typical variable capacitance diodes ranges from approximately 1 – 100 ns, depending primarily upon the base thickness and the carrier diffusion constant involved. Thus the frequency range over which the junction impedance is primarily resistive extends typically from dc up to an upper limit of about 1 – 100 MHz, depending upon the transit time. At much higher frequencies when the transit time is much *longer* than the period of the signal applied, both the

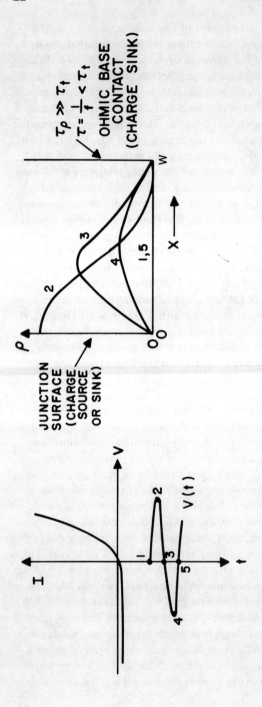

Fig. 2.4b Injected Minority Carrier Base Charge Density Distributions — Resultant Large Signal Base Distribution for the Applied Voltage Condition Shown

Physical Operation

conductive and reactive currents become equally important; i.e., $R_j \cong X_{Cj}$. In this high frequency range which extends from approximately 1 – 100 GHz and up for typical units, the condition of charge storage shown in Figure 2.4a is no longer correct. Under these conditions, the perturbation in charge density about the dc condition (indicated by the dotted lines) does not project all the way across the base region but indeed projects only a short distance from the junction or emitting surface thus grossly reducing the stored charge and the resultant diffusion capacitance. Furthermore, because of the time lag in the minority carriers diffusing in and out of the base region with respect to the applied voltage, the real and imaginary portions of the junction admittance become equal and increase as the square root of the frequency. Thus, in this frequency region, the capacitance and resistance representing the junction must both decrease as the square root of frequency. Between the low- and high-frequency regions discussed, there exists an intermediate frequency range in which the diffusion capacitance can dominate, depending upon the detailed doping profile in the base region. This intermediate frequency range, which is defined by the ratio of transit time to the period of the signal such that the condition $0.5 \leqslant \tau_t/\tau \geqslant 15$ exists, occurs between 5 MHz and 15 GHz for the typical range of base widths and associated transit times previously discussed. From the point of view of using a variable capacitance diode operating in the forward biased region, this frequency range where the diffusion capacitance dominates the impedance of the junction and is essentially independent of frequency is of the utmost importance.

The second form of capacitance, resulting from the injection capacitance, results from the storage of charge resulting from large signal excursions into the forward biased region. This effect is generally of interest only when the period of forward applied voltage is equal to or less than the transit time of minority carriers across the base region such that charge storage rather than rectification and its associated losses is obtained. To prevent rectification, the injected charge must not reach the ohmic contact nor recombine in the base region; i.e., $\tau < \tau_t$ and $\tau_\rho \gg \tau$. Thus for the cases of interest, only a transient base charge distribution is established with no equilibrium condition being reached. An example of such a transient, injected charge distribution existing in the base as a function of time is illustrated in Figure 2.4b for the situation in which a large signal, RF sine wave is applied at zero bias. The instantaneous charge distributions are shown for those instants of time noted on the sine wave for the situation in which there is no charge remaining in the base region at the start or completion of each cycle, such as the points labeled 1 and 5. At point 2, at the maximum of the forward going voltage, the maximum injection at the junction surface takes place. Since the transit time for injected carriers is of the same order as the period of RF some charge will just begin to reach the ohmic base contact or charge sink in this length of time as illustrated in the charge distribution labeled 2. As time progresses, the voltage decreases back to zero corresponding to point 3 and then goes negative such that the junction surface also becomes a charge sink or collector as well as that of the base contact. Under these conditions charge is rapidly drawn back across the reverse biased junction as well as to the ohmic contact which is represented by the distribution labeled 3. At the extreme of the negative or reverse going portion of the cycle, point 4, the charge in the base has significantly decreased from that of the previous distribution with its maximum density value now centered with respect to the reverse biased junction and the base contact. Finally, as the cycle is completed all the charge is exhausted from the base region and the cycle can be initiated again with zero history. The charge that is initially injected and then recovered during the reverse biased portion of the cycle represents an additional change in charge for a change in voltage, thus creating a greater effective junction capacitance than that attributable to the junction dipole layer alone, the additional increment being due to the injection capacitance. In utilizing this additional capacitance which is provided by the injected charge, it is important to maintain the period of forward bias short compared to the carrier transit time to minimize conductive currents and associated losses as previously noted. In this regard, it should be recognized that the transit time can be a function of the applied forward voltage, particularly for lightly doped base regions as used for high voltage varactors or PIN structures, as the motion of the carriers across the base is in part governed by the magnitude of the electric field created by the applied voltage. Thus, for a given RF frequency and bias, as the amplitude of the voltage is increased, the transit time will be shortened and eventually a voltage level will be reached when conduction currents will occur as carriers reach the ohmic base contact. Such conduction will cause rectification of the RF signal and a significant increase in the loss associated with the injection capacitance. This effect is represented by a sharp decrease in the value of shunt, junction resistance, R_j, shown in Figure 2.3. A further effect on the junction impedance is obtained if, for a given bias and RF voltage amplitude, the frequency of the applied voltage is increased or the period of forward applied voltage is decreased such that less charge enters the base region with a corresponding decrease in stored charge and resultant injection capacitance. Thus, at sufficiently high frequencies, the injection capacitance, like the small-signal diffusion capacitance, becomes insignificant in comparison with the depletion layer capacitance.

Before leaving the discussion of injection capacitance, it is important to point out two side effects associated with this capacitance which can prove to be undesirable in certain applications. First, as charge passes across the junction barrier either into or back out of the base region, shot noise is introduced. Thus to be complete, a noise current generator shunting the other junction impedance elements in the equivalent circuit of Figure 2.3 must be included. This additional source of noise present in the variable capacitance diode can be important in low noise systems

Physical Operation

which might employ these diodes in parametric amplifiers or in limiter components. The second effect which can result from injected carriers is a condition representative of a premature breakdown phenomena. As injected carriers are drawn back out of the base region with the reversal of applied junction voltage, they can be accelerated to such a point in passing through the junction as to produce additional carriers by a multiplication process, thereby incurring large reverse currents simulating a breakdown condition at voltages much less than the true breakdown voltage of the junction. This effect not only increases the losses by reducing the effective junction resistance, R_j, but can be extremely noisy and can lead to the catastrophic failure of the device due to the presence of high reverse currents sustained at high reverse voltages. This phenomenon definitely restricts the extent to which forward injection can be utilized, particularly in the situations in which large reverse voltage excursions will simultaneously take place.

2.2.2 Injection Losses and Base Conductivity Modulation

The losses associated with the total junction capacitance under forward biased operating conditions are essentially the same as those discussed for reverse biased operation except for the shunt junction resistance and the dependency of the base resistance primarily upon current rather than voltage. Thus the two series resistances representing the metallic and degenerate semiconductor resistances R_{ps} and R_{ns} in Figure 2.3 are the same as previously discussed for reverse biased operation, but the junction resistance, R_j, when under forward biased operating conditions is as significant a loss contribution to the variable capacitance as is the base resistance. As was discussed in the preceding section, this junction resistance can, at low frequencies, be the dominant junction impedance; whereas, at very high frequencies, it becomes comparable in magnitude to the diffusion capacitive reactance. Even under operating conditions when the diffusion or injection capacitance is the principal junction element, this shunt junction resistance will be as important as the base resistance in determining the Q of the variable capacitance diode under forward biased operating conditions. The base resistance, $R_b(I)$, in forward biased operation generally is significantly reduced by the action of the injected carriers; i.e., the presence of minority carriers in the base region causes a considerable increase in the bulk conductivity of this region, resulting in a marked reduction in this series resistive element as more current is passed. The extent of the change in base resistance with current depends on several factors of which the more important are the doping of the base region, the lifetime of carriers in this region and the width of the region with respect to the penetration of the injected carriers. The magnitude of this change for low voltage varactor elements typically ranges from no change to as much as a ten to one change with forward bias current; whereas, for extremely lightly doped base regions and high voltage diodes such as PIN structures, the change in base resistance from zero bias to a high forward bias (current density ≥ 100 A/sq. cm) can be as much as $10^4 - 10^5$ to 1. Thus the conductivity modulation effects in the base region resulting from forward biased operation can play a significant role in altering the total diode impedance or the Q associated with the variable junction capacitance. This phenomenon is extremely important in the operation of such diodes in limiter and switch functions if good component performance is to be achieved.

BIBLIOGRAPHY

Ghandhi, S.K., "The Theory and Practice of Microelectronics," J. Wiley and Sons, Inc., N.Y., N.Y. (1968).

Gibbons, J.F., "Semiconductor Electronics," McGraw-Hill Book Co., N.Y., N.Y. (1966).

Lindmayer J. and Wrigley, C.Y., "Fundamentals of Semiconductor Devices," D. Van Nostrand Co., Princeton, N.J. (1965).

Mattson, R.H., "Basic Junction Devices and Circuits," Wiley and Sons, Inc., N.Y., N.Y. (1963).

SMALL SIGNAL CHARACTERISTICS 3

3.0 General

With some understanding of the physical makeup and operation of the various variable capacitance diodes now in hand, the relationships for their small-signal characteristics will next be explored. In particular, relationships for the two primary properties of the semiconductor element, capacitance and resistance, will be obtained in terms of the device design (geometry and doping profile) and the appropriate material properties. Small-signal figures of merit for the devices are then presented which involve both the voltage dependent capacitance and its associated losses. The dependence of all of these properties on material choices, frequency of operation and temperature are discussed. The calculation of the capacitance and resistance for varactors with complex doping profile requires the use of numerical calculations as shown in the literature.[1]

3.1 Depletion Layer Capacitance

3.1.1 Capacitive Properties

The barrier capacitance, as described in the section on the physical operation of these devices, is that contribution to the total junction capacitance which results from the existence of the dipole or depletion layer produced by both the internal contact potential difference and any additional, externally applied potential. It represents almost the total junction capacitance when the diode is biased in the reverse state as no diffusion or injection capacitance then exists. The only additional contribution to the junction capacitance under such bias is that which can come from channeling and associated inversion layers near the peripheral surface of the junction. Such a contribution for well-fabricated diodes is generally small

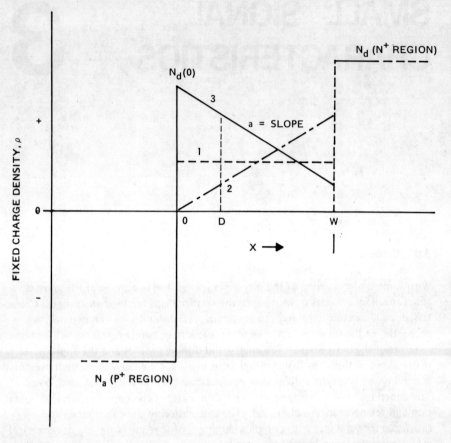

Fig. 3.1 Representative Fixed Charge Density Profiles for Analysis of the Depletion Layer Formation

in value, nonobservable at microwave frequencies because of isolating impedances and, at any rate, can usually be included by utilizing an extended junction area.

As the dipole layer, which constitutes the barrier capacitance, is totally governed by the charge density profile existing in the lattice through the junction and base regions, the doping profile produced in actual fabrication determines the barrier behavior with applied voltage. Three representative doping profiles which result in abrupt, linearly graded, and hyper-abrupt junctions will next be considered for analysis of their capacitive behavior.

Analysis of Depletion Layer and Resultant Capacitance: To illustrate the dipole or depletion layers and the barrier capacitances resulting from the junction doping

Small-Signal Characteristics

profiles discussed, consider the situation in which two opposite types of degenerate regions border on a central or base region of variable doping density. As shown in Figure 3.1, if the base region is arbitrarily selected to be N-type, then the three principal types of junction can be embraced by the following expression for the fixed charge distribution in the base region.

$$p(x) = e\, N_d(x) = e(N_d(0) + ax) \qquad 0 \leq x \leq W \qquad 3.1$$

where $p(x)$ is the fixed charge density distribution resulting from the base region doping (positive with ionized donors); e, the electronic charge; $N_d(x)$, the donor density; a, the slope or rate of change of donor density with distance through base x; and W, the total base thickness. The three most common junction types mentioned are indicated by the linear base doping or charge density profiles marked 1 through 3, where 1 represents the abrupt junction case, 2 the linearly graded case, and 3 the hyper-abrupt.

If now it is assumed that all mobile charge carriers (electrons) are swept out of the junction transition or depletion region (extending from $x = 0$ to $x = D$) as a result of the combined effects of the contact potential difference, ϕ_0, and the applied junction potential V_a, then Poisson's Equation (in one dimension) for the depletion region becomes

$$\frac{dE}{dx} = \frac{\rho(x)}{\epsilon} = \frac{e}{\epsilon}(N_d(0) + ax) \qquad 3.2$$

where E is the electric field in the depletion region and ϵ is the dielectric constant of semiconductor material employed. Integrating Equation 3.2 and recognizing that the field existing in the base outside the depletion region is small, so that $E(x \geq D) \approx 0$, the expression for the electric field within the depletion region becomes

$$E(x) = \frac{eN_d(0)}{\epsilon}(x - D) + \frac{ea}{2\epsilon}(x^2 - D^2), \qquad 0 \leq x \leq D \qquad 3.3$$

Integrating Equation 3.3, the electronic potential variation, $\phi(x)$, in the depletion region is obtained in terms of the assumed applied potential, V_a, occurring at $x=D$.

$$\phi(x) = V_a + \frac{eN_d(0)}{2\epsilon}(D-x)^2 + \frac{ea}{6\epsilon}(2D^3 - 3D^2 x + x^3), \qquad 0 \leq x \leq D \qquad 3.4$$

Further recognizing that $\phi(0) = \phi_0$, the contact potential difference occurring at the junction, the depletion layer thickness, D, can be expressed in terms of the doping profile and the total junction voltage, $V_j = (\phi_0 - V_a)$ (V_a considered positive for forward bias).

VARIABLE CAPACITANCE DIODES

$$D^3 + \frac{3N_d(0)}{2a} D^2 - \frac{3\epsilon}{ea} V_j = 0 \qquad 3.5$$

The two most common diode types are special cases of Equation 3.5. If a=0 such that uniform doping of the base region exists (Case 1 in Figure 3.1) the depletion layer thickness dependence on junction voltage for the abrupt junction is obtained.

$$D \text{ (abrupt junction)} = \left(\frac{2\epsilon}{eN_d(0)} V_j\right)^{1/2} \qquad 3.6$$

If a is assumed to be positive and $aD \gg N_d(0)$ (Case 2 in Figure 3.1), then the dependence for the linear graded junction is determined.

$$D \text{ (linear graded junction)} = \left(\frac{3\epsilon}{ea} V_j\right)^{1/3} \qquad 3.7$$

For these two cases, the barrier capacitance is simply obtained by recognizing the parallel plate capacitance equivalent of the dipole layer with effective thickness, D, dielectric constant ϵ, and cross-sectional or junction area A. Thus, the two most common junction capacitance relationships are given by

$$C \text{ (abrupt junction)} = \frac{\epsilon A}{D(V_j)} = \left(\frac{\epsilon e N_d(0)}{2}\right)^{1/2} \frac{A}{V_j^{1/2}} \quad \text{and} \qquad 3.8$$

$$C \text{ (linear graded junction)} = \frac{\epsilon A}{D(V_j)} = \left(\frac{\epsilon^2 ea}{3}\right)^{1/3} \frac{A}{V_j^{1/3}} \qquad 3.9$$

These two voltage-dependent capacitances are illustrated in Figure 3.2 by the normalized curves of the form

$$c \equiv C/C_o = (V_j V_{jo})^{-m} \equiv v_j^{-m} \qquad 3.10$$

where m indicates "the law," i.e., square or cube root for the two special, but common, cases cited.

Other depletion layer and barrier capacitance voltage dependencies can be obtained resulting in different "laws" depending on the doping profile. Consider the situation in which an abrupt change in doping takes place to create the junction (like the abrupt junction case) but where the doping or fixed charge density decreases with penetration into the base region (Case 3 in Figure 3.1) such that a is negative. This type of profile, which results in hyper-abrupt junctions, can be examined by normalizing the general expression for the depletion layer thickness and numerically evaluating the resultant, normalized capacitance-voltage relationship. If a D_o value is chosen such that

Small-Signal Characteristics

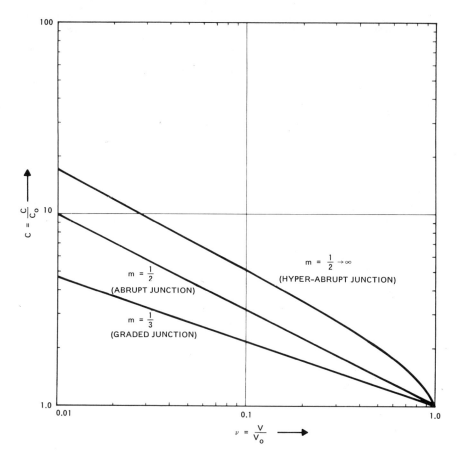

Fig. 3.2 Normalized Capacitance-Voltage Relationships for Three Types of Junction

$$|a| \, D_0 = N_d(0) \text{ for some } V_{j0} \qquad 3.11$$

and it is recognized that with a negative value of a the sign of the charge density is not to change (creation of a second junction), i.e., $|a|W \leqslant N_d(0)$, a normalized expression for the depletion layer dependence is obtained from 3.5.

$$D^2 (D - 1.5 \, D_0) = -\frac{3\epsilon}{ea} V_j \quad (a \text{ assumed to be negative}) \qquad 3.12$$

With $V_j = V_{j0}$ specified when $D = D_0$, then

$$D_0^3 = \frac{6\epsilon}{ea} V_{j0} \qquad 3.13$$

Dividing both sides of 3.12 by D_0^3 and defining $d \equiv \dfrac{D}{D_0}$, the normalized expression is obtained.

$$d^2(d - 1.5) = -0.5 \, v_j \qquad 3.14$$

This relationship, together with the fact that $c = 1/d$ permits the evaluation of a representative case of the hyper-abrupt junction, voltage-dependent capacitance as indicated Figure 3.2. As can be seen this type of junction possesses a range of voltage dependencies ($m = \frac{1}{2}$ to ∞) or "laws" depending on the bias point chosen. Thus it is possible to obtain with the three doping profiles considered nearly the entire possible range of capacitance-voltage laws. It might also be noted that although the doping or fixed charge density variations chosen here for illustrative purposes are linear, considerable curvature can exist in these profiles (as can occur on actual fabrication) without appreciable deviation from the laws obtained.

It must be recognized that the barrier capacitance-voltage relationships obtained are bounded with respect to voltage for actual devices. In reverse bias operation for which these relationships are most useful, the voltage range is ultimately limited by breakdown. This limit is reached when the field anywhere within the junction reaches the critical value, E_i, required for ionization and avalanche multiplication. From 3.3 it can be determined that the maximum field occurs at $x = 0$, thus

$$E_{max} = E(0) = -\frac{eN_d(0)}{\epsilon} D - \frac{ea}{2\epsilon} D^2 \qquad 3.15$$

If E_{max} is set equal to E_i and substitution for D from 3.5 is made, expressions for the breakdown voltage can be obtained. For example, from 3.6 and 3.7, the expressions for the breakdown voltage for the abrupt and graded junctions are obtained as follows:

$$V_{jb} \text{ (abrupt junction)} = \frac{\epsilon E_i^2}{2eN_d(0)} \qquad 3.16$$

$$V_{jb} \text{ (linear graded junction)} = \left(\frac{8\epsilon E_i^3}{9ea}\right)^{1/2} \qquad 3.17$$

Of course it must be noted here that the breakdown voltage of actual variable capacitance diodes can be less than the bulk junction values determined by 3.16 and 3.17 because of surface phenomena.

In addition to the breakdown voltage, another reverse bias voltage value can be of importance, namely the punch-thru voltage or the voltage at which the depletion

Small-Signal Characteristics

layer, D, equals the base thickness W. To be a condition of practical concern, the punch-thru voltage must occur prior to breakdown with increasing reverse bias. This condition is generally only reached for diodes with light base doping (small value of $N_d(0)$ and/or a) which results in large depletion layer thicknesses or for very thin-base diodes. The punch-thru voltage can be determined by equating the expressions for depletion layer thicknesses (3.6 and 3.7) to the base thickness, W. Thus, the punch-thru voltages V_{jp} become

$$V_{jp}(\text{abrupt junction}) = \frac{eN_d(0)W^2}{2\epsilon} \qquad 3.18$$

$$V_{jp}(\text{linear graded junction}) = \frac{ea\,W^3}{3\epsilon} \qquad 3.19$$

Beyond this voltage until breakdown is reached, the capacitance becomes essentially *independent* of the applied voltage and 3.8, 3.9, and 3.14 become invalid in this voltage range.

The barrier capacitance relationships are also bounded in the forward bias direction. Ideally, when the applied voltage equals the contact potential difference such that $V_j = 0$, the depletion layer thickness goes to zero and the barrier capacitance becomes infinite. However, under these same conditions, the current across the junction would become infinite, preventing V_j from approaching zero because of the series resistance. In practice, the barrier capacitance is generally a *small* contributor to the total junction capacitance when appreciable forward current ($V_a \approx \phi_0/2$) and associated charge injection take place as the diffusion or injection capacitance is then dominant. The exception to this situation occurs only in units possessing extremely short lifetimes compared with the transit time of the injected charge such that the effective diffusion capacitance remains small compared to the barrier capacitance. It should be noted here that the barrier capacitance expressions for small excursions into the forward bias region ($I_{fwd} < 10 - 100\ I_s$) are quite useful as representative of the total junction capacitance, particularly for high frequency applications where the diffusion capacitance contributions diminish.

Variation of C with Frequency: The frequency dependence of the depletion layer capacitance, C, is determined by the time required to move the majority carriers in increasing or decreasing the width, D, of the depletion layer. This time lag is determined by the time constant, R_sC, of the series circuit composed of the depletion layer capacitance, C, and the series resistance, R_s, of the varactor. Although this time lag is evident from circuit considerations, it is also possible to derive it by looking directly at the motion of the majority carriers in the device. If, at t=0, a voltage V is applied to the device, this voltage will produce an electric field, E = V/W, where W is the thickness of the resistive part of the varactor. This electric field will cause the depletion layer to change in width with a velocity, v_d,

given by
$$v_d = \mu E = \mu V/W \qquad 3.20$$

where μ is the mobility of the majority carriers. If the initial thickness of the depletion layer is D_0 and the final value is D, the time it takes to establish steady state conditions is

$$t = \frac{(D-D_0)}{v_d} = \frac{(D-D_0)}{\mu V} W \qquad 3.21$$

assuming v_d is constant for this equivalent time interval. If the number of impurities is N (all assumed ionized), then it follows from Poisson's equation that

$$\frac{V}{D} = \frac{eN(D-D_0)}{\epsilon} \quad . \text{ Therefore,} \qquad 3.22$$

$$t = \frac{\epsilon W}{eN\mu D} = \frac{W}{eN\mu A} \frac{\epsilon A}{D} = R_s C \qquad 3.23$$

In the above derivation, it has been assumed that the majority carriers with a velocity, $v_d = \mu E$, move in phase with the applied electric field. This is the case for times longer than the mean collision time which, in semiconductors, is of the order of 10^{-13} sec.

As mentioned in Chapter II, the speed of response, or the R_sC time constant for typical varactors, is in the range of 10^{-10} to 10^{-13} seconds, which corresponds to frequencies in the millimeter wave range.

Variation of C with Temperature: As shown in Equations 3.8 and 3.9, the depletion layer or barrier capacitance of a varactor is given by

$$C = \frac{\epsilon A}{D} \qquad 3.24$$

where D, the thickness of the depletion layer, is determined by

(1) the junction potential V_j across the junction and equal to

$V_j = \varphi_0(T) - V_a$ (φ_0 and ϕ_0 used interchangeably for contact potential) and

(2) the ionized impurity concentration in the base of the device, N(T).

The temperature variation of C, for a fixed applied voltage V_a, is caused by the

Small-Signal Characteristics

temperature variations of either the contact potential φ_0 or the number of ionized impurities, since the dependence of the permittivity, ϵ, upon temperature over the normal operating range is negligible.

a) Effect of $\varphi_0(T)$ - (V_a and $N(T)$ held constant). The temperature dependence of the capacitance, C, on the contact potential difference, $\varphi_0(T)$, can be determined from the equations

$$C = B V_j^{-m} \qquad 3.25$$

$$V_j = \varphi_0(T) - V_a \qquad 3.26$$

$$\varphi_0(T) = \frac{kT}{e} \ln \frac{N_a N_d}{N_i^2} \qquad 3.27$$

For small ΔT and for fixed N_a, N_d and V_a, it is shown that

$$\frac{\Delta C}{C} = \frac{-m}{V_j} \left(\varphi_0 - \frac{\mathcal{E}_g}{e} + \frac{T}{e} \frac{d\mathcal{E}_g}{dT} - \frac{3kT}{e} \right) \frac{\Delta T}{T} \qquad 3.28$$

where $\mathcal{E}_g$ is the energy gap of the semiconductor involved. Assuming $\mathcal{E}_g$ to have a linear temperature dependence $\mathcal{E}_g = \mathcal{E}_g(0) + AT$, where $\mathcal{E}_g(0)$ is the energy gap at $0°K$, it follows that

$$\frac{\Delta C}{C} = \frac{-m}{V_j} \left(\varphi_0 - \frac{\mathcal{E}_g(0)}{e} - \frac{3kT}{e} \right) \frac{\Delta T}{T} \qquad 3.29$$

Since φ_0 is always less than $\frac{\mathcal{E}_g}{e}$, V_j always positive and m always positive, it follows that ΔC is positive for a positive ΔT. An order of magnitude evaluation of the above equation for $\varphi_0 \approx 1/4 \, \mathcal{E}_g(0)/e$, $V_j = \varphi_0$, $T=300°K$ and $\Delta T=1°C$ gives

$$\frac{\Delta C}{C} = 0.33\% \text{ per } °C.$$

For large temperature variations, the variation of C with temperature has to be calculated using Equations 3.25 - 3.27. Useful equations for determining this dependence are the following:

$$\frac{C(T)}{C(T_0)} = \left[\frac{\varphi_0(T) - V_a}{\varphi_0(T_0) - V_a} \right]^{-m} \qquad 3.30$$

$$\frac{\varphi_0(T)}{\varphi_0(T_0)} = \frac{T}{T_0} \left[1 - \left(\frac{3kT_0}{e\varphi_0(T_0)} \right) \ln \frac{T}{T_0} - \frac{\mathcal{E}_g(0)}{e\varphi_0(T_0)} \left(1 - \frac{T_0}{T} \right) \right] \qquad 3.31$$

where $C(T)$ and $\varphi_0(T)$ are respectively, the capacitance and contact potential difference at a temperature T. Figures 3.3 and 3.4 show the measured variation of C with temperature for two different silicon varactors at two different bias voltages, and the evaluated points using Equations 3.30 and 3.31. For this comparison, the values of φ_0 used in the calculated curves were determined from the best match of measured and theoretical curves for $V_a = 0$. The resulting values of φ_0 appear to correspond well with those which would be expected for the breakdown voltages of these units. As can be seen from the graphs, the high voltage varactors, due to the low value of ϕ_0 with its associated temperature sensitivity, exhibit a larger temperature variation of capacitance, particularly at zero bias. As would be expected, a much reduced temperature variation is obtained with the varactor biased in the reverse direction with $V_a \geqslant \phi_0$.

b) Effect of $N(T)$. The thickness of the depletion layer depends upon the number of ionized impurities in the base of the device. For a fixed junction voltage voltage, the smaller the number of ionized impurities the larger the thickness of the depletion layer and hence, the smaller the capacitance. Although it is possible to determine from Equations 3.8 and 3.9 the change in capacitance with the number of ionized impurities, it is important to realize that the change in the number of ionized impurities occurs, in most cases, over a small temperature range. Figure 3.5 shows the variation of the fraction N/N_d of ionized donors for N-type silicon with temperature, calculated from the equation [2]

$$\frac{kT}{\mathcal{E}_d} = \frac{1}{\ln \dfrac{N_c}{2N_d} + \ln \dfrac{N_d}{N}\left(\dfrac{N_d}{N} - 1\right)} \qquad 3.32$$

where N_c is the effective density of states of the conduction band and $\mathcal{E}_d$ is the donor ionization energy. The transition temperature T_t at which $N = N_d/2$ is given by

$$T_t = \frac{\mathcal{E}_d/k}{\ln \dfrac{N_c}{N_d}} \qquad 3.33$$

Near this transition temperature small changes in temperature result in large changes in the number of ionized donors. Note, however, that this temperature typically occurs well below room temperature. As $N(T)$ affects $\phi_0(T)$, both its direct and indirect influence on the capacitance must be accounted for in evaluating this dependence.

Small-Signal Characteristics

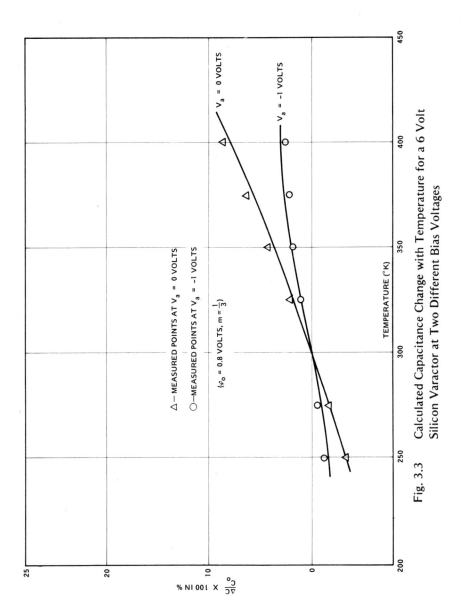

Fig. 3.3 Calculated Capacitance Change with Temperature for a 6 Volt Silicon Varactor at Two Different Bias Voltages

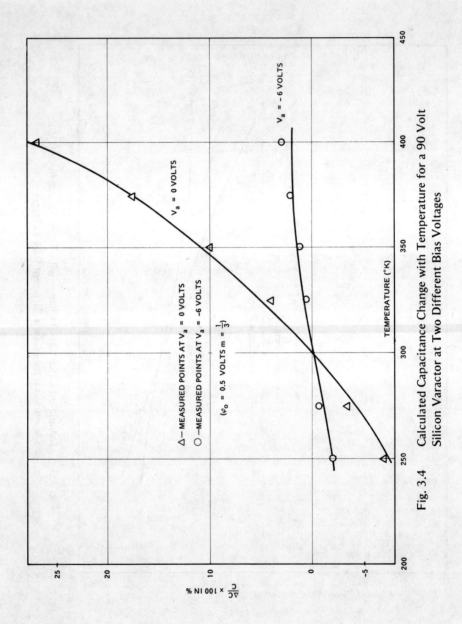

Fig. 3.4 Calculated Capacitance Change with Temperature for a 90 Volt Silicon Varactor at Two Different Bias Voltages

Small-Signal Characteristics

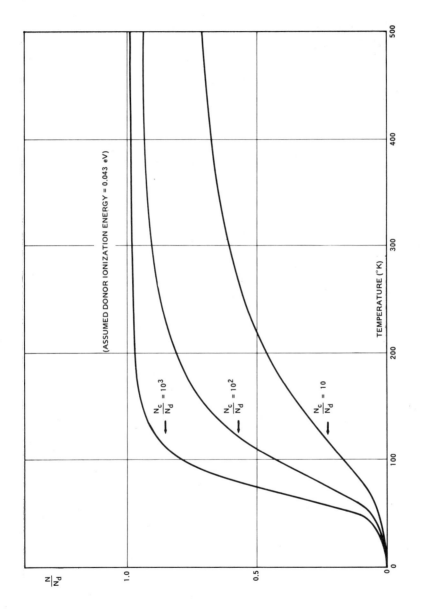

Fig. 3.5 Calculated Fraction of Ionized Donors for N-type Silicon

Variation of C with Material Parameters: The main dependence of C upon materials is through the permittivity, ϵ, and the contact potential difference, φ_0. As far as the permittivity is concerned, there is not much choice in the case of Ge, Si and GaAs, since all of them have the same bonding arrangement and have similar dielectric constants. These are given in the following table:

	Ge	Si	GaAs
ϵ/ϵ_0	16.3	11.8	12

The contact potential difference of a varactor can be controlled by an appropriate choice of the impurity doping levels of the P and N regions. However, the contact potential is always less than the energy band gap which makes it possible to have a larger range of contact potential differences in the wider gap semiconductors.

3.1.2 Associated Losses

The ohmic losses of a varactor are caused by a leakage resistance R_j across the junction and the total series resistance R_s of the diode.

The leakage resistance R_j is mainly due to surface leakage currents and is very much dependent upon the passivation process chosen by the manufacturer in the fabrication of the varactor. It is a resistance that does not generally play an important role in varactors due to the high frequency use of the device. That is, the frequency of operation typically is such that $\frac{1}{\omega C} \ll R_j$, which makes the current through the junction mainly a displacement current and not a conductive current.

The series resistance R_s consists, in the case of a P^+NN^+ varactor, of the resistance R_b of the lightly doped N region, of the resistances R_{ps} and R_{ns} of the P^+ and N^+ regions, and of the package resistance R_{pack}. Of the above resistances, the most important one is the resistance R_b contributed by the base of the diode. A discussion of the evaluation of R_b and of its frequency and temperature dependence is given next.

Evaluation of R_b: In the case of a P^+NN^+ (or N^+PP^+) varactor, the resistance R_b is voltage-dependent because of the penetration of the depletion layer into the base region. Thus, as the depletion layer changes with applied voltage, the remaining or resistive portion of the base must change with voltage. The expression for this resistance is given by

$$R_b = \frac{1}{Ae} \int_D^W \frac{dx}{\mu N(x)} \qquad 3.34$$

Small-Signal Characteristics

where $N(x)$ is the number of ionized impurities. Although the mobility, μ, is concentration dependent, the variation of μ in the base region can be neglected. Assuming $N(x)$ to be of the form $N(x) = N(0) + ax$, it follows that

$$R_b = \frac{1}{Ae\mu a} \ln\left[\frac{N(0) + aW}{N(0) + aD}\right] \qquad 3.35$$

The above equation can be normalized with respect to a resistance, R_{bw}, equivalent to the base resistance for a homogeneous carrier concentration of $N(0) + aW$, given by

$$R_{bw} = \frac{W}{Ae\mu[N(0) + aW]} \qquad 3.36$$

Thus, equation 3.35 can be rewritten as

$$r = \frac{R_b}{R_{bw}} = \left[1 + \frac{N(0)}{aW}\right] \ln\left[\frac{1 + \frac{aW}{N(0)}}{1 + \frac{aD}{N(0)}}\right] \qquad 3.37$$

Introducing the quantity D_0 defined by Equation 3.11, the above equation changes to

$$r = \left[1 \pm \frac{D_0}{W}\right] \ln\left[\frac{1 \pm \frac{W}{D_0}}{1 \pm \frac{D}{D_0}}\right] \quad (\text{– sign applies when } a \text{ is negative}) \qquad 3.38$$

For the case of an abrupt junction, $a = 0$ and $D_0 = \infty$ which gives

$$r_{abrupt} = 1 - \frac{D}{W} \qquad 3.39$$

For a linearly graded junction, $N(0) = 0$ and $D_0 = 0$ which gives

$$r_{graded} = \ln\frac{W}{D} \qquad 3.40$$

Since the thickness D of the depletion layer depends upon the potential V_j applied across the junction, these resistance expressions, Equations 3.39 and 3.40 are voltage-dependent. Figures 3.6 and 3.7 are plots of these resistance values as a function of the normalized voltage V_j/V_{jp}, where V_{jp} is the voltage for the punch-thru

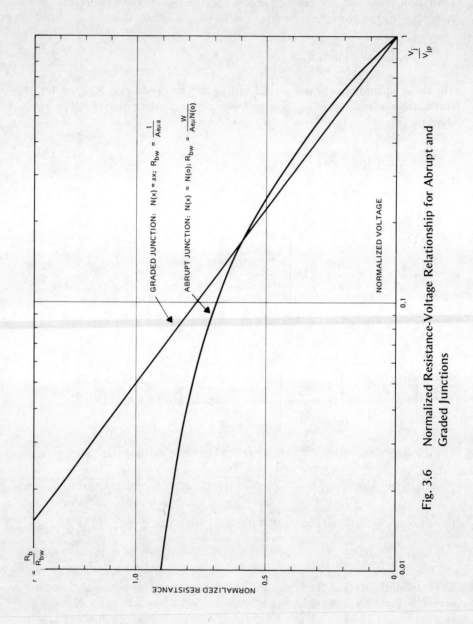

Fig. 3.6 Normalized Resistance-Voltage Relationship for Abrupt and Graded Junctions

Small-Signal Characteristics

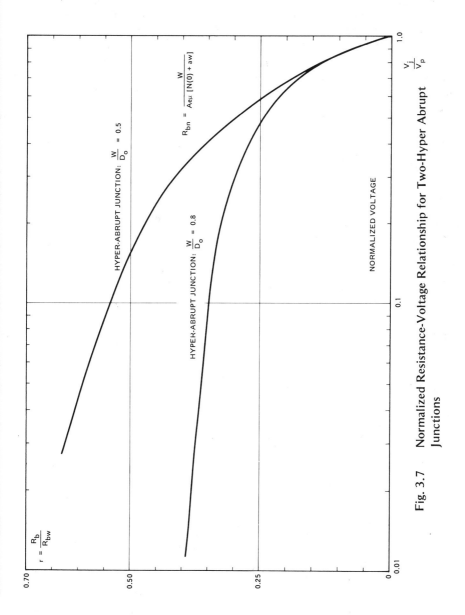

Fig. 3.7 Normalized Resistance-Voltage Relationship for Two-Hyper Abrupt Junctions

condition. The data for the plots in Figure 3.6 are obtained from Equations 3.39 and 3.40 utilizing Equations 3.6, 3.7, 3.18 and 3.19 to express D/W in terms of of V_j/V_{jp}. The data for the plots in Figure 3.7 are obtained from Equation 3.38 with assumed values of W/D_0 and using Equation 3.14 to determine D/D_0 for values of V_j/V_{jo} which in turn are converted to values of V_j/V_{jp} by using Equation 3.14 again to gain the factor V_{jo}/V_{jp} from the assumed values of W/D_0 (letting D=W). These curves show that the larger the reverse voltage applied across the junction, the smaller the value of the base resistance, R_b. Although the hyper-abrupt junctions seem to display less variation of R_b with voltage, this is due to the inherently large series resistance built into the base.

Frequency Dependence of R_b: At low frequencies the current through the base of a varactor is simply determined by the base resistance, R_b, and the voltage across the base. At high frequencies two effects can change this: the displacement current in the base region and the skin effects which increase R_b.

The frequency, ω_r, at which the displacement current has the same magnitude as the conduction current is called the dielectric relaxation frequency and it is given by

$$\omega_r = \frac{\sigma}{\epsilon} = \frac{1}{\rho \epsilon} \qquad 3.41$$

For silicon varactors with breakdown voltages less than 100 volts, the resistivity, ρ, of the base is 1 ohm-cm or less, such that ω_r is 10^{12} rad/sec. or higher which falls in the millimeter wave range.

The most important factor to be taken into account in the calculation of R_b is the magnitude of the skin depth compared to the dimensions of the base region. The skin depth, δ, as a function of frequency and resistivity, ρ, is given by

$$\delta \text{ (in mils)} = 62.7 \sqrt{\rho/f} \qquad 3.42$$

where ρ is expressed ohm-cm and f in GHz. For resistivities less than 1 ohm-cm and for frequencies larger than 1 GHz, the skin effect has an important role in determining the value of R_b in large area varactors.

Temperature Dependence of R_b: The temperature dependence of R_b is contained in the temperature dependence of the resistivity, ρ, of the base material. The resistivity ρ is inversely proportional to the mobility, μ, and the carrier concentration, n. The mobility has a complicated temperature dependence since it is determined by a combination of lattice scattering and impurity scattering. In general, for varactor applications, it is more meaningful to present data of ρ versus T rather than to split ρ into μ and n. Figures 3.8 and 3.9 show the temperature dependence of $\sigma (= 1/\rho)$

Small-Signal Characteristics

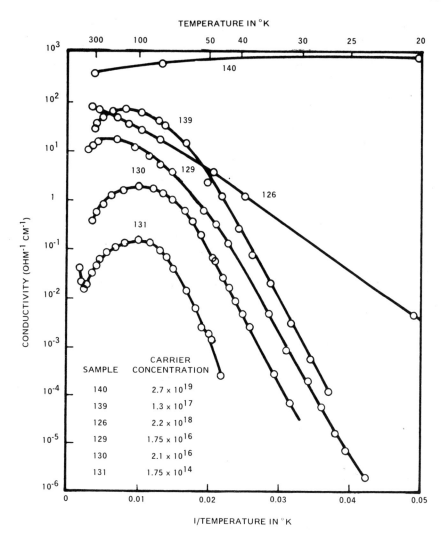

Fig. 3.8 Temperature Dependence of Electrical Conductivity for Arsenic-Doped Silicon Samples

for n-type silicon and n-type gallium arsenide with different impurity concentration. For low resistivity material, the resistivity is almost temperature independent. For high resistivity material, the resistivity increases with temperature below the intrinsic range and decreases with temperature above the intrinsic range.

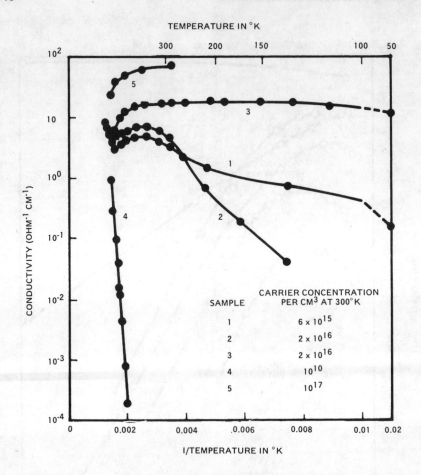

Fig. 3.9   Temperature Dependence of Electric Conductivity for N-type Gallium Arsenide Samples

3.1.3 Q-Value and Cut-off Frequency

The small-signal equivalent circuit of a varactor is, neglecting lead inductance and package capacitance, the parallel combination of the junction capacitance C and the leakage resistance R_j, in series with the total series resistance, R_s of the diode. At low frequencies such that $\frac{1}{\omega C} \approx R_j \gg R_s$, the Q of the varactor, defined as the ratio of twice the average energy stored to the energy dissipated per cycle, is given by

$$Q = \omega C\, R_j \qquad \qquad 3.43$$

which, for constant C and R_j, increases in direct proportion with the circular frequency ω.

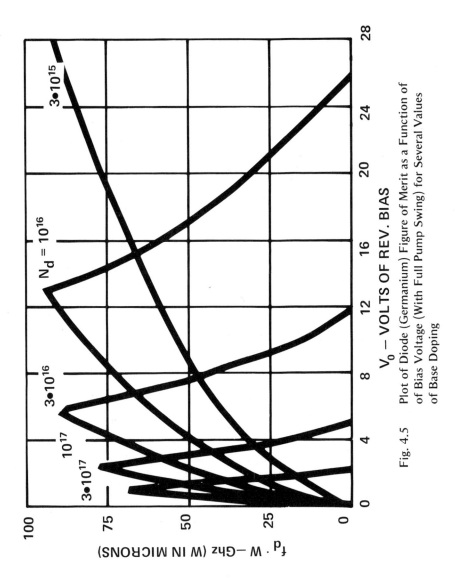

Fig. 4.5 Plot of Diode (Germanium) Figure of Merit as a Function of Bias Voltage (With Full Pump Swing) for Several Values of Base Doping

Large-Signal Characteristics

forward bias voltage V_{fwd}. V_{fwd} is the maximum forward bias permitted before shot noise becomes significant and depends upon several factors including frequency of operation, base thickness, and doping. For the purposes of this evaluation, however, a constant value of 0.1 V has been assumed for V_{fwd}. The values of a_n/a_0^2 used are taken from Figure 4.3.

From Figure 4.5, it can be seen that for each choice of base doping, as the bias (and thus also the pump voltage) is increased from zero, the diode figure of merit rapidly increases until a maximum value is reached at a voltage corresponding to $1/2\ (|V_b|-|V_{fwd}|)$. If the bias is still further increased, the value of the figure of merit rapidly drops off, becoming zero at the breakdown voltage. The explanation for these "saw-tooth"-like curves can be readily gained from Equation 4.14, for, assuming fixed geometry and doping, f_d is proportional to the relative pump swing and the bias voltage. Thus near zero bias both V_0 and γ (and thus a_n/a_0^2) are small; at $1/2\ (|V_b|-|V_{fwd}|)$, V_0 is approximately half its maximum value while γ (and a_n/a_0^2) is at a maximum and dominates; near breakdown, V_0 is maximum but γ (and a_n/a_0^2) is approaching zero and again dominates.

From the dependency of f_d upon bias, it was concluded that to obtain the maximum possible figure of merit the bias must be set at $1/2\ (|V_b|-|V_{fwd}|)$. A plot of f_d (max) vs. doping is presented in Figure 4.6, both for the case of direct pumping n = 1, and subharmonic pumping n = 2 and n = 3 (no resonant mode at two or three times the pump frequency, i.e., no harmonic *power* generated).

It is observed that for all three conditions of pumping, f_d(max) can be increased by reducing the doping (at least over the range $N_d = 3 \cdot 10^{15} - 10^{18}$ atoms/cc). The reason for the more rapid increase in f_d(max) with n = 2 and 3 is that in all cases the dependency upon doping is almost entirely through γ and a_n/a_0^2, and a_2/a_0^2 and a_3/a_0^2 are faster functions of γ than a_1/a_0^2. (See Figure 4.3.) It should be noted here that, if V_{fwd} had not been assumed constant at 0.1 V, but had been assumed to increase somewhat with doping, then the maximum relative swing and therefore a_n/a_0^2 and f_d(max) would be higher than that shown for heavier doping. However, in general, f_d(max) would still be decreased by increasing the base doping.

In many instances, a compromise in amplifier or converter performance is or must be made so as to reduce the pump power or dissipation requirements at some sacrifice in f_d. For example, the amplitude of the voltage required for direct pumping to yield f_d(max) with a base doping of $3 \cdot 10^{17}$ atoms/cc in germanium is 1.1 volts as compared to 5.8 volts for $3 \cdot 10^{16}$ atoms/cc. (See Figure 4.6.) Thus, for a reduction in pump power of 28/1 (proportional to the square of the pump amplitudes), a reduction in f_d(max) of only 26% is involved. To

Using this definition of the diode figure of merit, it is immediately possible to determine the minimum amplifier noise figure for a chosen signal frequency by recognizing that the minimum value for the quantity within brackets of Equation 4.10 is obtained when $f_2 = f_d$. Thus,

$$F_{min} \cong 1 + (2 f_1/f_d) \qquad 4.13$$

with the idle frequency chosen to be equal to the diode figure of merit.

It can be noted here that, since f_d can be evaluated for any voltage-dependent capacitor (knowing its C-V and R-V laws and choice of bias and relative pump swing), this choice of diode figure of merit is directly useful in comparing capacitors of various types as well as optimizing their respective designs. Furthermore, since this figure of merit is directly involved in the choice of idle frequency, it determines the amplifier noise figure for a specified signal frequency. Such a figure of merit should prove very useful to the amplifier circuit designer as well.

Evaluation of f_d: To determine the behavior of this figure of merit with material, device, and operating parameters, an evaluation of Equation 4.11 must be made. For the abrupt junction case, the evaluation is most readily made by using Equation 4.11c which defines the figure in terms of the operating point cutoff frequency, f_{co} which is given by Equation 3.53 for the condition when $R_s \approx R_b$ and $D \ll W$ ($b_0 \approx 1$).

$$f_d \text{ (abrupt junction)} = \frac{\mu}{\pi} \left(\frac{eN}{8\epsilon}\right)^{1/2} \frac{(\phi_0 + V_0)^{1/2}}{W} \frac{a_n(\gamma)}{a_0^2(\gamma)} \qquad 4.14$$

with $V_j = (\phi_0 + V_0)$. In a similar manner using Equation 4.11a and the expression for C_0 and $R_s \approx R_b$ with $b_0 \approx 1$ (Equations 3.9, 3.40 and 3.36), the diode figure of merit for the linearly graded junction can be obtained.

$$f_d \text{(graded junction)} = \frac{\mu}{\pi} \left(\frac{\sqrt{3}\, ea}{8\epsilon}\right)^{2/3} \frac{(\phi_0 + V_0)^{1/3}}{\ln\frac{W}{D}} \frac{a_n(\gamma)}{a_0^2(\gamma)} \qquad 4.15$$

From an examination of Equations 4.14 and 4.15, it can be seen that f_d depends upon choice of material, doping, bias, and relative pump swing. For the purpose of detailed examination, the variation of f_d with choice of V_0 using the maximum permissible swing (γ) for various values of doping has been plotted in Figure 4.5 for germanium abrupt junction varactors. The maximum permissible swing is determined by the ratio $V_m/(\phi + V_0)$, where V_m (pump voltage amplitude) is limited by the choice of V_0, the breakdown voltage V_b, and/or the maximum

Thus, the effective large-signal base resistance can be as much as 24% greater than the small-signal value at the operating bias point. This difference can appreciably reduce the effective varactor Q and thus the projected noise figure performance of a parametric amplifier.

4.1.3 Time-Dependent Based Figure of Merit

As the time-dependent characterization of the varactor is most useful in the large-signal (pump), small-signal (information) applications such as parametric amplifiers or converters, it is appropriate to derive a device figure of merit for these instances. Using the time-dependent representation of the varactor, as described previously, it is possible to derive an expression for the noise figure of a parametric amplifier from which the definition of a diode figure of merit follows directly. One such noise figure expression for high gain amplifier operation under conditions where the input circuit contributes little noise compared to the source is as follows[1]

$$F = 1 + [(f_1/f_2) + (f_1 f_2/f_d^2)] \qquad 4.10$$

where f_1 = signal frequency, f_2 = idle frequency ($=f_3 - f_1$ for the regenerative amplifier with f_3 = pump frequency), and

$$f_d \equiv a_n/4\pi a_0^2 C_0 R_s \qquad 4.11a$$

which is termed the diode figure of merit and has the units of frequency. C_0 is the small-signal value of capacitance at the operating bias point as defined previously together with the Fourier coefficients, a_n. R_s is the average value of diode series resistance at the operating bias point. If nearly all of the resistance of the diode occurs in the base region such that resistance is also time-dependent as described in the previous section, then

$$R_s \approx R_b = b_0 R_0 \qquad 4.12$$

and the diode figure of merit becomes

$$f_d \approx a_n/4\pi a_0^2 b_0 C_0 R_0 \qquad 4.11b$$

This figure can in turn be related to the cutoff frequency as follows

$$f_d \approx f_{co} (a_n/2a_0^2 b_0) \qquad 4.11c$$

where f_{co} = cutoff frequency at the operating bias point = $(2\pi C_0 R_0)^{-1}$.

with $\theta = \omega_p t$, n = harmonic number, $R_0 = R_b(V'_0) = R_{bw} r(V'_0)$ = the base resistance at the bias point from Equation 3.36 and 3.38, and $r[V'(\theta)]$ = the normalized resistance given by Equation 3.38 with the substitution made for $D(V')$ with $V' = V_0 + \phi_0 + v_p(\theta)$.

The resistance coefficients, b_n, can be evaluated for the abrupt junction varactor by noting that

$$r[V'(\theta)]_{abrupt} = 1 - \frac{D[V'(\theta)]}{W} = 1 - K_w \left[\frac{V'(\theta)}{V'_0}\right]^{1/2} \qquad 4.6$$

where $K_w = D(V'_0)/W$ = constant for a given bias = $\left[\frac{2\epsilon V'_0}{eN_d(0)W^2}\right]^{1/2}$ from Equations 3.39 and 3.6. Substituting Equation 4.6 into 4.5 and again defining $\gamma = V_m/V'_0$, the relative pump swing, the integral form of the coefficient expression becomes

$$b_n(abrupt) = \frac{2}{\pi(1 - K_w)} \int_0^\pi [1 - K_w(1 + \gamma \cos\theta)^{1/2}] \cos n\theta \, d\theta, \qquad 4.7$$

(exception: $b_0 = 1/2$ this value)

In a manner similar to that employed in the evaluation of the capacitive coefficients, the resistive coefficients, b_n, can be found in terms of the complete elliptic integrals of the first and second kind and k as previously defined under Equation 4.3. The resulting first or average value coefficient is given by

$$b_0(abrupt) = \frac{1}{(1 - K_w)} - \frac{2K_w}{\pi(1 - K_w)} (1 + \gamma)^{1/2} E(k) \qquad 4.8$$

If desired, additional, higher order coefficients of possible importance in the determination of conversion gain can be calculated from Equation 4.7 for the abrupt junction case. To determine the deviation of b_0, the normalized average value from one, consider the case when $V'_0 = V'_p/2$ for the full pump swing such that $D(V'_0) = W/\sqrt{2}$ from Equation 3.18 and $K_w = 1/\sqrt{2}$. Then

$$b_0 (abrupt, V'_0 = V'_p/2) = 3.414 - 1.536(1 + \gamma)^{1/2} E(k) \qquad 4.9a$$

or with $\gamma \approx 1$, $k = 1$, and $E(1) = 1$,

$$b_0(abrupt, V'_0 = V'_p/2, \gamma \approx 1) = 1.24 \qquad 4.9b$$

Large-Signal Characteristics

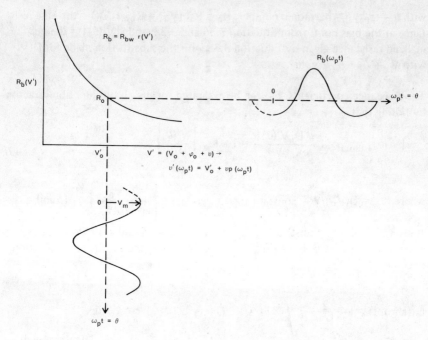

Fig. 4.4 Production of a Time-Dependent Resistance by Pumping a Voltage-Dependent Resistance

Shown in Figure 4.4 is a R-V characteristic representative of either an abrupt or graded junction varactor. This voltage-dependent base resistance is simply the general normalized value, r, from Equation 3.38 multiplied by the constant R_{bw} given by Equation 3.36. The voltage dependence of this resistance is, of course, derived from the dependence of D, the depletion layer thickness, on the total applied voltage, V', as obtained from Equations 3.5, 3.6, and 3.7. With an assumed sinusoidal pump voltage applied, the voltage-dependent base resistance becomes a time-dependent element subject to Fourier analysis as done for the time-dependent capacitance. Such an analysis yields the following series representation for the resistance:[1]

$$R_b(\omega_p t) = b_0 R_0 + b_1 R_0 \cos \omega_p t + b_2 R_0 \cos 2\omega_p t \cdots + b_n R_0 \cos n\omega_p t \quad 4.4$$

where b_n is given by

$$b_n = \frac{2}{\pi} \int_0^\pi \frac{r[V'(\theta)]}{r[V_0']} \cos n\theta \, d\theta \quad \text{(exception: } b_0 = 1/2 \text{ this value)} \quad 4.5$$

VARIABLE CAPACITANCE DIODES

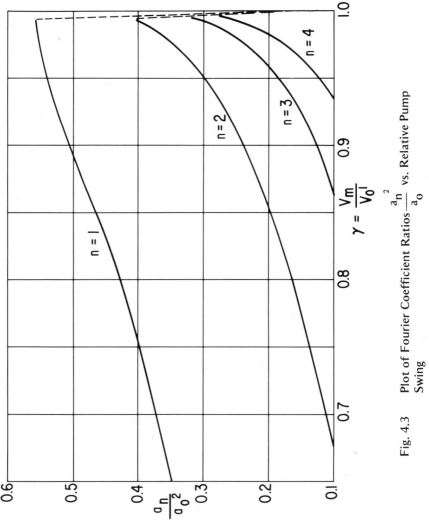

Fig. 4.3 Plot of Fourier Coefficient Ratios $\dfrac{a_n}{a_o}$ vs. Relative Pump Swing

Large-Signal Characteristics

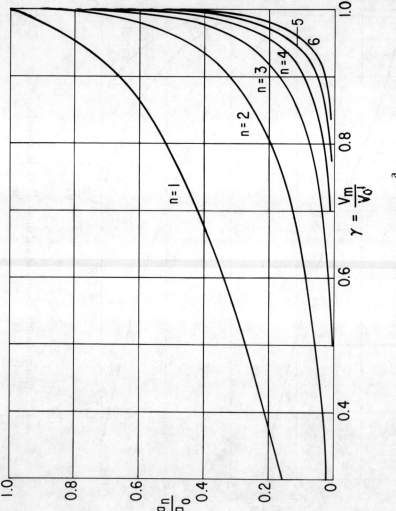

Fig. 4.2 Plot of Fourier Coefficient Ratios $\frac{a_n}{a_o}$ vs. Relative Pump Swing

where K = complete elliptic integral of the first kind, E = complete elliptic integral of the second kind, and $k^2 = 2\gamma/1 + \gamma$. The numerical values of K and E as a function of k^2 (a function of the relative pump swing) are given in most mathematical handbooks or tables or can be determined using a computer program. Values of a_n vs. γ have been obtained and the significant ratios a_n/a_0 and a_n/a_0^2 vs. γ are plotted in Figures 4.2 and 4.3, respectively. The ratio a_n/a_0 is important in the amplifier gain expression in that the negative conductance generated is proportional to the square of this ratio in typical operation. It can be observed that a_n/a_0 has significant values for the higher harmonics as compared to the fundamental ratio provided "hard" pumping ($\gamma > 0.9$) is employed. The ratio a_n/a_0^2 appears in the amplifier high gain noise figure expression and should, in general, be as large as possible. From Figure 4.3, it can be seen that a maximum value of this ratio a_n/a_0^2 occurs for values of γ approaching unity as both a_n and a_0 approach infinity.

Approximate values for the graded junction case can be obtained from the abrupt case if it is recognized that the values of a_n ($n > 0$) and the change in a_0 (i.e., $a_0 - 1$) are approximately 36 per cent less when $m = 1/3$ as compared to $m = 1/2$ for sinusoidal pumping.

It should be noted that other, more general determinations of the capacitance coefficients can be made by numerical evaluation of Equation 4.2 for any value of m. An equally valid characterization of the varactor for these operating constraints can also be obtained from an equivalent elastance or charge vs. voltage representation. Some parametric operations have been analyzed using these forms of large-signal characterization.[3]

4.1.2 Time-Dependent Base Resistance

For high-quality varactors, i.e., those with high Q or f_c values, considerable variation in the base resistance can occur with the application of reverse bias voltage (see Section 3.1.2 and Figure 3.6). Thus, just as the capacitance can become time-dependent by the action of a pumping voltage, so also can the base resistance. This large-signal behavior of the base resistance can be of considerable importance in the accurate projection of parametric amplifier or converter operation. At sufficiently high frequencies such that $Q_0 \approx 10$, the variation in the varactor resistance for large pump swing is sufficient to significantly modify the effective Q-values so as to lower the anticipated noise performance or conversion gain.

Large-Signal Characteristics

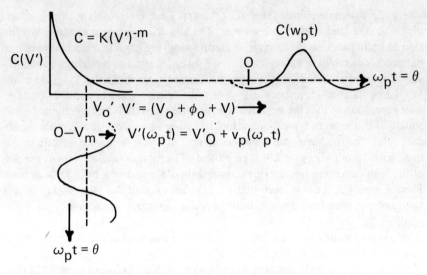

Fig. 4.1 The Production of a Time-Dependent Capacitance by Pumping a Voltage Dependent Capacitance

The capacitance coefficients, a_n, can be evaluated for the abrupt junction varactor, $m = 1/2$, by the following procedure. Replace $\cos \theta$ and $\cos n\theta$ in Equation 4.2 by $(2 \cos^2 n\theta/2 - 1)$ breaking the single integral into $(n + 1)$ integrals. By letting $k^2 = 2\gamma/1 + \gamma$, $\alpha = \theta/2$, and utilizing the relationship $\cos^2 \alpha = 1 - \sin^2 \alpha$, each of the integrals in turn can be expressed in the standard form of complete elliptic integrals of the first and second kind and elliptic functions. The resulting coefficients are as follows:

$$a_0 = \frac{\sqrt{2}}{\pi\sqrt{\gamma}} \, k \, K,$$

$$a_1 = -\frac{2\sqrt{2}}{\pi\sqrt{\gamma}} \left[(k - 2/k) K + 2/k \cdot E \right]$$

$$\tag{4.3}$$

$$a_2 = \frac{2\sqrt{2}}{\pi\sqrt{\gamma}} \left[(k - 16/3k + 16/3k^3) K + (8/3k - 16/3k^3) E \right],$$

$$a_3 = -\frac{2\sqrt{2}}{\pi\sqrt{\gamma}} \left[(k - 158/15k + 128/5k^3 - 256/15k^5) K + (46/15k - 256/15k^3 + 256/15k^5) E \right], \text{ etc.,}$$

tivity modulation). The resulting two-state representation, a small RF resistance for forward bias and a high Q, small capacitance for the reverse bias, permits relatively simple component analysis and design.

The above three forms of large-signal characterizations will be described in the following sections. In addition, figures of merit useful for varactor and PIN selection in large-signal applications are presented.

4.1 Varactor as a Time-Dependent Element

4.1.1 Time-Dependent Capacitance

If for the particular application at hand the voltage swing is primarily restricted to the reverse bias region as in low noise parametric operation, then the total capacitance of the varactor element is essentially just that contributed by the depletion layer. This condition, which neglects any contribution from injection or diffusion capacitance, permits the varactor to be represented by a time-dependent capacitance expressed in the form of a Fourier series which is valid up to frequencies whose period approaches the basic time constant of the device.(See Equation 3.23.)

Depicted in Figure 4.1 is a C-V characteristic representative of either the abrupt or graded junction varactor (m = 1/2 or 1/3, respectively in Equation 3.10). Impressed upon this characteristic is a bias voltage, V_0' (including the internal diode voltage ϕ_0), and an assumed sinusoidal pump voltage, v_p ($\omega_p t$) whose amplitude is V_m. The resulting time-dependent capacitance waveform, $C(\omega_p t)$, is shown to the right of the characteristic. A Fourier analysis of this capacitance waveform yields the desired harmonic time-dependent capacitances which can be utilized for amplifier or converter operation. Such an analysis yields the following series representation for the capacitance[1,2]:

$$C(\omega_p t) = a_0 C_0 + a_1 C_0 \cos \omega_p t + a_2 C_0 \cos 2\omega_p t \cdots + a_n C_0 \cos n\omega_p t \qquad 4.1$$

where a_n is given by

$$a_n = \frac{2}{\pi} \int_0^\pi \frac{\cos n\theta \, d\theta}{(1 + \gamma \cos \theta)^m} \quad \text{(exception: } a_0 = 1/2 \text{ this value)} \qquad 4.2$$

with $\theta = \omega_p t$, n = harmonic number, $\gamma = V_m/V_0'$, the relative pump swing, m = exponent of the C-V law, and C_0 = the capacitance at the bias point. It should be noted that the capacitance coefficients depend *only* upon the C-V law and *relative* pump swing.

LARGE SIGNAL CHARACTERISTICS   4

4.0   General

In many applications of varactors the large-signal or nonlinear characteristics are utilized. For example, in both parametric amplifier and harmonic generator applications, a large voltage signal must be applied to the varactor to achieve the desired mixing action or frequency multiplication. Because of the difficulty in identifying the varactor characteristics for a significant frequency range and because of large voltage swings which include excursions into both the reverse and forward bias regions, no one unique characterization exists which is useful. In addition, the usual nonlinear dependence of the capacitance or charge storage on voltage further complicates any characterization or analysis.

Generally, the two most useful approaches to the large-signal characterization problem for conventional variable capacitance diodes are the concepts of time-dependent capacitance (or charge) and current-determined charge storage. The first concept, time-dependent capacitance, is most useful in parametric amplifier and converter design where the pumping operation (voltage swing) is usually confined to the reverse bias region to insure low noise performance. The second concept, current-determined charge storage, is of most value in determining high frequency harmonic generator performance for which any large-signal waveform can be applied to the diode which acts as a charge storage tank dependent upon the current flow.

For the varactor or PIN diode being used as a switching element, a two-state characterization is generally most useful. In this case, the forward and reverse bias states are identified as an essentially voltage independent capacitor with and without the shunting effect of the base-or I-region conductive plasma (or conduc-

REFERENCES

1. T.P. Lee, "Calculations of Cutoff Frequency, Breakdown Voltage and Capacitance for Diffused Junctions in their Epitaxial Silicon Layers," *IEEE Trans. on Elec. Dev.*, ED-13, 881-896 (December 1966).

2. J. Lindmayer and C. Y. Wrigley, "Fundamentals of Semiconductor Devices," D. VanNostrand Co., Princeton, N. J. (1965).

3. M. Uenohara, "Cooled Varactor Parametric Amplifiers," in *Advances in Microwaves*, 2, ed. L. Young, Academic Press, Inc., N.Y., N.Y. (1967)

4. R.I. Harrison, "Parametric Diode Q Measurements," *Microwave Journal*, Vol. 3, 43-46 (May 1960).

5. W. Schockley, "The Theory of p-n Junctions in Semiconductors and p-n Junction Transistors," *Bell Syst. Tech. J.* 28, 435-489 (1949).

BIBLIOGRAPHY

Dickens, L.E., "Spreading Resistance as a Function of Frequency," *IEEE Trans. Microwave Theory and Techniques*, MTT 15, 101-108 (February 1967).

Grayzel, A. I., "The Cutoff Frequency of a Varactor Diode with Variable Series Resistance," *Proc. IEEE* (correspondence) 54, 875 (June 1966).

Irvin, J.C., Lee, T.P., Decker, D.R., "Varactor Diodes in Microwave Semiconductor Devices and their Circuit Applications," ed. H.A. Watson, McGraw-Hill Book Co., N.Y., N.Y. (1969).

Kennedy, D.P., "Spreading Resistance in Cylindrical Semiconductor Devices," *J. of App. Phys.*, 31, 1490-1497 (August 1960).

Lee, T.P., "Evaluation of Voltage Dependent Series Resistance of Epitaxial Varactors," *IEEE Trans. on Elec. Dev.*, ED-12, 457-470 (August 1965).

McMahon M.E., and Straube, G.F., "Voltage-Sensitive Semiconductor Capacitor," *IRE WESCON Conv. Rec.*, Rec., pt. 3, 72-82 (August 1958).

Morgan, S.P., and Smits, F.M., "Potential Distribution and Capacitance of a Graded p-n Junction," *B.S.T.J.* 39, 1573-1602 (November 1960).

Mortenson, K.E., "Parametric Diode Figure of Merit and Optimization," *J. of App. Phys.*, 31, 1207-1212 (July 1960).

Penfield, P. Jr., "Maximum Cutoff Frequency of Varactor Diodes," *Proc. IEEE* (correspondence) 53, 422-423 (April 1965).

Sze, S.M., "Physics of Semiconductor Devices," John Wiley and Sons, N.Y., N.Y. (1969).

Vendlin, G.D., "Dependence of Varactor Cutoff Frequency on Extrinsic Series Resistance and Bias Voltage," *Proc. IEEE* (correspondence) 54, 54-55 (January 1966).

Small-Signal Characteristics

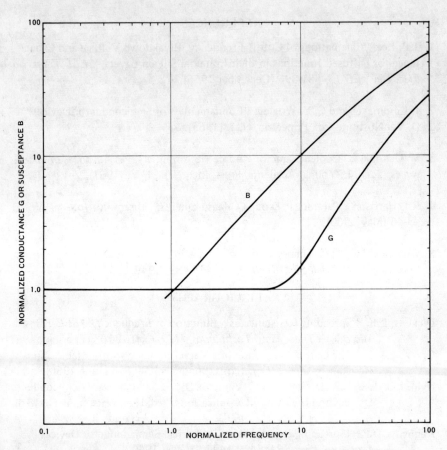

Fig. 3.15 Normalized Plot of the Conductance G and Susceptance B of a P^+NN^+ Diode as a Function of $\omega\tau_p$ for the Case $\frac{L_p}{L_r} = 10$; $\frac{W}{L_p} = 1$

which shows that $Q > 1$ for $\omega\tau_p > 1$.

Figure 3.15 shows the plot of Equation 3.73 for the case $W/L_p = 1$, $L_p/L_r = 10$. The above plot shows that $Q = 1$ occurs at $\omega\tau_p \approx 1$.

$$Y_p = \frac{e}{kT}(J_p+J_{ps}) \frac{\dfrac{\sqrt{1+\left(\dfrac{L_p}{2L_r}\right)^2}+i\omega\tau_p}{\tanh\dfrac{W}{L_p}\sqrt{1+\left(\dfrac{L_p}{2L_r}\right)^2+i\omega\tau_p}} - \left(\dfrac{L_p}{2L_r}\right)}{\dfrac{\sqrt{1+\left(\dfrac{L_p}{2L_r}\right)^2}}{\tanh\dfrac{W}{L_p}\sqrt{1+\left(\dfrac{L_p}{2L_r}\right)^2}} - \left(\dfrac{L_p}{2L_r}\right)} \qquad 3.72b$$

where J_p is the dc bias current and J_{ps} is the leakage current of the varactor. The admittance, Y_p, can be normalized with respect to the admittance at zero frequency, $\frac{e}{kT}(J_p + J_{ps})$. If $\omega\tau_p \gg 1$, $\left(\frac{L_p}{2L_r}\right)^2$ and $\left(\frac{W}{L_p}\right)^2$, then the normalized admittance, y_p, is given by

$$y_p = \frac{Y_p}{\frac{e}{kT}(J_p+J_{ps})} \approx \frac{\sqrt{\dfrac{\omega\tau_p}{2}}(1+i) - \left(\dfrac{L_p}{2L_r}\right)}{\dfrac{\sqrt{1+\left(\dfrac{L_p}{2L_r}\right)^2}}{\tanh\dfrac{W}{L_p}\sqrt{1+\left(\dfrac{L_p}{2L_r}\right)^2}} - \left(\dfrac{L_p}{2L_r}\right)} \qquad 3.73$$

which shows that for high enough frequencies, the susceptance is larger than the conductance. That is, the Q of the diode is greater than one. This result is to be noted in comparison with that obtained for the case of uniform base diodes, where the Q is always less than one. The reason the susceptance is larger than the conductance ($Q > 1$) in this case is that the built-in electric field applies a force to the injected carriers opposing their diffusing to the end of the base where they disappear by recombination. This charge retention effect increases the charge stored and the susceptance while reducing the conduction process, thus yielding the improvement in Q. It should be emphasized that the built-in electric field is effective only so long as the injection level is small. At high injection levels the built-in electric field is masked by that created due to the injected carriers and is thus ineffective in influencing the motion of carriers.

The minimum value of $\omega\tau_p$ at which the Q is larger than one can be determined as follows. If $L_p/2L_r \gg 1$, then

$$y_p = 1 + i\omega\tau_p \qquad 3.74$$

Small-Signal Characteristics

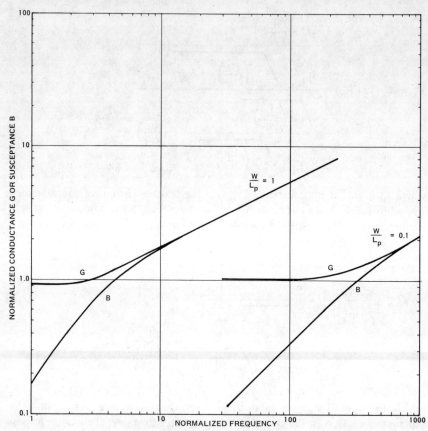

Fig. 3.14 Normalized Plot of the Conductance G and Susceptance B of P^+NN^+ Diodes as a Function of $\omega\tau_p$ for $\frac{W}{L_p} = 1$ and $\frac{W}{L_p} = 0.1$

$$Y_p = \frac{e\mu_p\, p_n}{L_p} e^{\frac{eV_0}{kT}} \left[\frac{\sqrt{1 + \left(\frac{L_p}{2L_r}\right)^2} + i\omega\tau_p}{\tanh\frac{W}{L_p} \sqrt{1 + \left(\frac{L_p}{2L_r}\right)^2 + i\omega\tau_p}} - \left(\frac{L_p}{2L_r}\right) \right] \quad 3.72a$$

which can be written as

Similarly, for an infinitely long diode, $W/L_p \gg 1$, the admittance becomes

$$Y_p \approx \left(\frac{eJ_p}{kT} - \frac{W}{L_p}\right)\sqrt{1 + i\omega\tau_p} \qquad 3.70$$

where now J_p is given by

$$J_p = \frac{ep_n D_p}{L_p} e^{\frac{eV_p}{kT}} \qquad 3.71$$

It should be noted that Equations 3.69 and 3.70 are exactly the same, as they should be since for $\omega\tau_p \gg 1$, independently of W/L_p, the diode should behave as an infinitely long diode.

The dependence of Y_p upon bias current, temperature and lifetime (or diffusion length) of minority carriers is evident from Equation 3.67. The admittance is directly proportional to the forward bias current and inversely proportional to temperature. Some additional temperature dependence can also exist through the temperature dependence of the diffusion constant and the lifetime. The main effect of the lifetime upon the admittance is to determine the frequency at which the diffusion capacitance starts to be of importance. The smaller the lifetime the higher the frequency at which the capacitive susceptance (charge storage effect) starts to become appreciable compared to the conductance.

Figure 3.14 shows a plot of the real and imaginary parts of the admittance expression given by Equation 3.67 as functions of $\omega\tau_p$ for two values of W/L_p in order to illustrate the dependence of the admittance upon geometry. It is seen that the main effect of the W/L_p ratio (with regard to the $\omega\tau_p$ dependence) is to determine the value of $\omega\tau_p$ at which storage effects start to take place. In addition, note that the admittance plots of Figure 3.14 have been normalized with respect to the admittance at zero frequency, which for the two cases shown can be very different.

3.2.3 Capacitance and Conductance with Graded Base

In the case of a graded base varactor with a retarding electric field, the admittance per unit area from Equation 3.64 is given by

Small-Signal Characteristics

3.2.2. Capacitance and Conductance with Uniform Base

For the case of a uniform base diode $L_r = \infty$ and the expression for the admittance per unit area becomes

$$Y_p = \frac{e p_n e^{\frac{eV_0}{kT}} \mu_p}{W} \left[\frac{\left(\frac{W}{L_p}\right) \sqrt{1 + i\omega\tau_p}}{\tanh\left(\frac{W}{L_p}\right)\sqrt{1+i\omega\tau_p}} \right] \qquad 3.65$$

The term outside the brackets is related to the bias current, J_p (assuming the saturation or leakage current to be very small), and can be written

$$e \frac{p_n}{W} e^{\frac{eV_0}{kT}} \mu_p \approx J_p \frac{e}{kT}, \text{ if } \frac{W}{L_p} \ll 1 \text{ and } \tanh \frac{W}{L_p} \approx \frac{W}{L_p}, \text{ or,} \qquad 3.66$$

$$Y_p \approx \frac{eJ_p}{kT} \frac{\frac{W}{L_p}\sqrt{1 + i\omega\tau_p}}{\tanh \frac{W}{L_p} \sqrt{1+i\omega\tau_p}} \qquad 3.67$$

If the hyperbolic cotangent is expanded in a series and only the first two terms taken into account, the expression for Y_p becomes

$$Y_p \approx \frac{eJ_p}{kT} \left[1 + \frac{1}{3}\left(\frac{W}{L_p}\right)^2 (1 + i\omega\tau_p) \right] \qquad 3.68a$$

which is valid if $\left| \frac{W}{L_p}\sqrt{1 + i\omega\tau_p} \right| < 1$ (diffusion length long compared to the base width and the ac period correspondingly no shorter than the lifetime). If, in addition to the above condition, it is also true that $\omega\tau_p > 1$ (ac period shorter than the lifetime), then

$$Y_p \approx \frac{eJ_p}{kT}\left[1 + \frac{i}{3}\left(\frac{W}{L_p}\right)^2 \omega\tau_p \right] \qquad 3.68b$$

From Equation 3.67, if $\left|\frac{W}{L_p}\sqrt{1 + i\omega\tau_p}\right| \gg 1$ (diffusion length short compared to the base width and/or the ac period short compared to the lifetime), then the admittance becomes

$$Y_p \approx \left(\frac{eJ_p}{kT} \frac{W}{L_p}\right)\sqrt{1 + i\omega\tau_p} \qquad 3.69$$

$$p - p_n = p_1 \approx p_n e^{\frac{v_1}{kT}} e^{\frac{eV_0}{kT}} e^{i\omega t} \qquad 3.59$$

where v_1 is the small ac incremental voltage applied in series with the dc bias voltage V_0. The boundary condition at $x = W$, that is, at the boundary between the N and the N^+ region must also be determined. First, it is recognized that an NN^+ junction is permeable to the flow of majority carriers (electrons) and impermeable to the flow of minority carriers. That is, the minority carriers try to accumulate at an NN^+ boundary. The extent of this accumulation is a function of the recombination conditions or minority carrier lifetime in the N^+ region. With short lifetime (the usual situation), the conditions are such that no accumulation takes place and the boundary condition at $x = W$ is expressed by

$$p - p_n = 0 \qquad 3.60$$

With the above boundary conditions, the minority carrier concentration is given by

$$p - p_n = -\frac{p_1 e^{-\beta W} e^{(-\alpha+\beta)x}}{2 \sinh \beta W} + \frac{p_1 e^{\beta W} e^{-(\alpha+\beta)x}}{2 \sinh \beta W} \qquad 3.61$$

The current density (combining Equations 3.55 and 3.56) expressed by

$$J_p = -eD_p \left[\frac{\partial p}{\partial x} + \frac{p}{L_r} \right] \qquad 3.62$$

yields an ac component by substitution of Equation 3.61 at $x = 0$, given by

$$j_p \Big|_{x=0} = eD_p p_1 \left[\beta \frac{\cosh \beta W}{\sinh \beta W} - \alpha \right] \qquad 3.63$$

The admittance per unit area, j_p/v_1, is found by combining Equation 3.59 with Equation 3.63 to yield

$$Y_p = e p_n e^{\frac{eV_0}{kT}} \mu_p \left[\beta \frac{\cosh \beta W}{\sinh \beta W} - \alpha \right] \qquad 3.64$$

It should be mentioned that the above admittance expression is only valid in the case of no accumulation of minority carriers at the NN^+ junction and under low level injection conditions. At high injection levels, the built-in electric field due to any impurity concentration is reduced effectively to zero.

Small-Signal Characteristics

barrier structure and the first case corresponds to the so-called P-N junction which is the case generally found in P^+NN^+ varactors. The flow of excess holes in the base is mainly by diffusion or by diffusion and drift in the case of a built-in electric field as occurs in graded base structures.

In this section, the expressions for the small signal diffusion capacitance and conductance of abrupt and graded P^+NN^+ structures are derived.[5]

The equations which determine the transport of the excess minority carriers are the continuity equation (one-dimensional)

$$\frac{\partial p}{\partial t} = -\frac{p-p_n}{\tau_p} - \frac{1}{e}\frac{\partial J_p}{\partial x} \qquad 3.54$$

and the current equation

$$J_p = -eD_p \frac{\partial p}{\partial x} - e\mu_p p \frac{\partial \psi}{\partial x} \qquad 3.55$$

where ψ is a retarding potential in the base created by an increasing donor concentration in the base region

$$\frac{\partial \psi}{\partial x} = \frac{kT}{e}\frac{1}{N_d}\frac{\partial N_d}{\partial x} = \frac{kT}{eL_r} \qquad 3.56$$

where L_r is the distance through which ψ increases $\frac{kT}{e}$ or $N_d(x)$ increases by 63.6%. For the case of constant electric field and sinusoidal time variation of the hole concentration in the base, the following ac component equation is obtained

$$\frac{\partial^2 p}{\partial x^2} + \frac{1}{L_r}\frac{\partial p}{\partial x} - \frac{(1+i\omega\tau_p)}{L_p^2}(p-p_n) = 0 \qquad 3.57$$

where $L_p^2 = D_p \tau_p$. The solutions of the above equation are of the form

$$p - p_n = Ae^{(-\alpha+\beta)x} + Be^{-(\alpha+\beta)x}, \text{ where} \qquad 3.58$$

$$\alpha = \frac{1}{2L_r} \quad \beta = \frac{1}{L_p}\sqrt{\left(\frac{L_p}{2L_r}\right)^2 + (1+i\omega\tau_p)}$$

The values of the constants A and B are determined by the boundary conditions.

In the case of a junction where the current is limited by the transport of the minority carriers in the base and not by their supply from the P^+ region, the boundary condition at $x = 0$ is given by

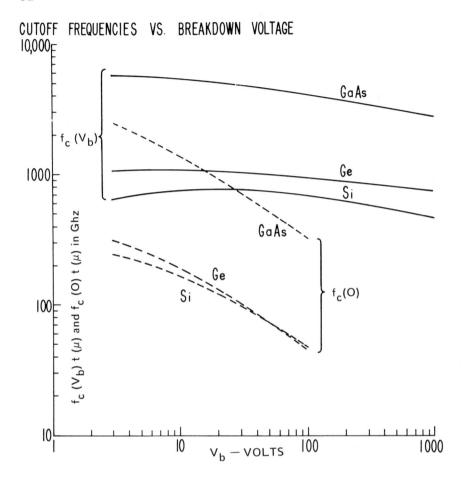

Fig. 3.13   Cutoff Frequencies vs. Breakdown Voltage

3.2.1 General Admittance Properties

When a forward voltage is applied across the terminals of a P^+NN^+ varactor, voltage is divided among the several regions of the varactor. Part of this voltage appears across the P^+ and N^+ regions where the conduction of current is mainly by drift of majority carriers. The rest of the voltage appears mainly across the P^+N depletion layer since no depletion layer is formed at the NN^+ junction. The forward voltage across the P^+N depletion layer produces injection of holes into the base. The current flow is determined by either the flow of the injected excess holes in the base or by majority carrier (electrons) passage from the base to the emitter. The latter case corresponds to a Schottky

Small-Signal Characteristics

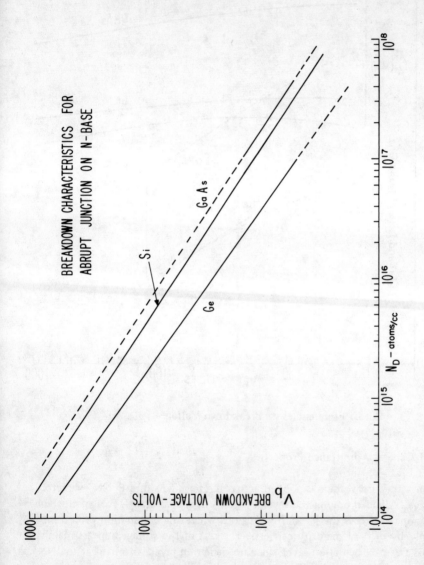

Fig. 3.12 Breakdown Characteristics for Abrupt Junction on N-Type Base

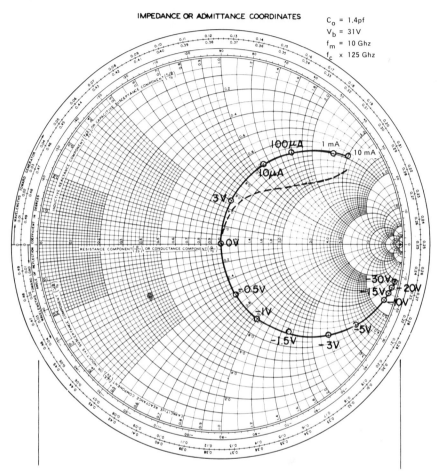

Fig. 3.11 Smith Chart Plot of GaAs Varactor Impedance

concentration, N, or the breakdown voltage V_b. Figure 3.12 is a plot of breakdown voltage versus impurity concentration for Ge, Si and GaAs. Figure 3.13 shows $f_c(V_b)$ and $f_c(0)$ for Ge, Si and GaAs varactors as functions of the breakdown voltage. A very important conclusion from Figure 3.13 is that $f_c(V_b)$ is almost independent of V_b. This is due to the fact that, as shown in Figure 3.12, V_b is almost inversely proportional to N. The difference in $f_c(V_b)$, for the above three materials, is caused by the difference in the values of $\mu/\epsilon^{1/2}$ for the three materials.

3.2   Injection or Diffusion Capacitance

Small-Signal Characteristics

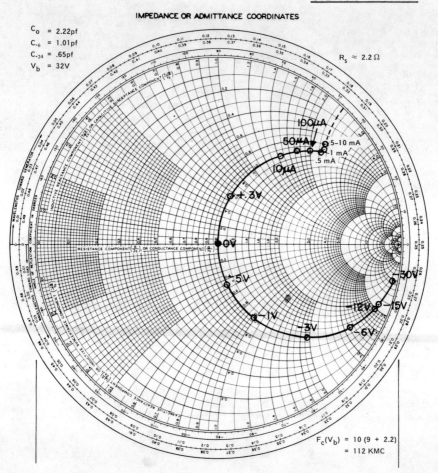

Fig. 3.10 Smith Chart Plot of Silicon Varactor Impedance

$$D = \left(\frac{2\epsilon}{eN} V_j\right)^{1/2} \quad\quad 3.52$$

Substitution of Equation 3.52 into 3.51 gives

$$f_c W = \frac{\mu}{\pi} \left(\frac{e}{2\epsilon}\right)^{1/2} (NV_j)^{1/2} \quad\quad 3.53$$

In the application of varactors, the cut-off frequency at breakdown, $f_c(V_b)$, and at zero bias, $f_c(0)$, are two very important parameters. From Equation 3.53, it follows that $f_c(V_b)$ and $f_c(0)$ are, for a given material, a function of the impurity

$$\frac{z(V)}{r_s} = z(V) = 1 - i\,[Q_c(V) - Q_c(0)] = 1 - iQ(V) \qquad 3.48$$

The true Q_c- value of the varactor can be determined at any reverse bias point by observing the normalized reactance value from the Chart (in essence $Q(V)$) and adding $Q_c(0)$ to it. For example, the true varactor Q at breakdown is

$$Q_c(V_b) = Q(V_b) + Q_c(0) \qquad 3.49$$

The value of $Q_c(0)$ can be obtained by noting the value of Q for high forward bias where $C \to \infty$ and $Q_c \to 0$ and no further reactance change with voltage takes place. This point is readily noted for silicon units as R_s decreases because of the conductivity modulation produced by the injected charge at high forward bias. The cut-off frequency of the varactor can be simply stated from the Q value determined in Equation 3.49 and knowledge of the measurement frequency, f_0. Thus

$$f_c(V) = f_0 Q_c(V) \qquad 3.50$$

Figures 3.10 and 3.11 show the normalized impedance of a Si and a GaAs varactor measured at 10 GHz as a function of voltage under matched conditions at zero bias. The above figures show that the measured points follow Equation 3.49. Figure 3.10 shows that conductivity modulation (reduction in R_s) does take place, for the particular varactor, at currents higher than 0.5 mA. From such data, it is determined that the $Q_c(0)$ of the varactor is 2.2, $Q(V_b)$ is 9, $Q_c(V_b)$ is $(9 + 2.2) = 11.2$ and that f_c is 112 GHz. The data taken on the GaAs varactor, shown in Figure 3.11, does not show any conductivity modulation effects. This is understandable since the excess carrier lifetime in GaAs is much smaller than the excess carrier lifetime in silicon. The low value of excess carrier lifetime is also thought to explain some anomalous data as shown in Figure 3.11 as a dotted line. Such an apparent increase in the resistance can be explained if the shunting effect of rectification takes place under forward bias conditions. For rectification to take place, it is necessary to have a small value of lifetime, which is the case with GaAs.

Dependence of Cut-Off Frequency on Materials: The cut-off frequency of a varactor is defined by Equation 3.45. For the case of an abrupt P^+NN^+ varactor, the value of R_s can be taken as the resistance of the base region. In this case, the cut-off frequency is given by

$$f_c = \frac{e\mu ND}{2\pi\epsilon(W - D)} \approx \frac{e\mu ND}{2\pi\epsilon W} \qquad 3.51$$

The value of D, the thickness of the depletion layer, depends upon the junction voltage and it is given by

Small-Signal Characteristics

At high frequencies, such that $R_j \gg \frac{1}{\omega C} \approx R_s$, the Q of the varactor is

$$Q = \frac{1}{\omega C R_s} \qquad 3.44$$

which decreases with frequency.

Equation 3.44 is usually taken as the definition of the Q of varactors, since the frequency at which they are generally used makes Equation 3.44 valid.

The cut-off frequency f_c of a varactor is defined as the frequency at which Q = 1. That is

$$f_c = \frac{1}{2\pi C R_s} \qquad 3.45$$

It should be pointed out that f_c, as well as Q, are functions of the voltage V_j applied across the junction. The variation of Q with voltage at high frequencies is usually represented in a Smith chart. The basis of the representation is as follows. The impedance of the varactor at a frequency ω is given by

$$Z(V) = R_s - \frac{i}{\omega C(V)} \qquad 3.46$$

In varactor measurements using the match method[4] it is customary to match the varactor to the line at zero bias. In this manner, the impedance of the varactor becomes normalized to the line or guide impedance such that at zero bias the series resistance becomes unity and the net reactance zero. Thus, the normalized, matched varactor impedance becomes

$$z(V) = r_s - \frac{i}{\omega Z_0}\left[\frac{1}{C(V)} - \frac{1}{C(0)}\right] \text{ or,} \qquad 3.47a$$

$$= r_s - \frac{i}{\omega C(0) Z_0}\left[\frac{C(0)}{C(V)} - 1\right] \qquad 3.47b$$

where the net reactance at zero bias equals zero and $r_s(0) = \frac{R_s}{Z_0} = 1$.

A plot of $z(V)$ is, in Smith chart form, a circle of constant resistance ($r_s = 1$). For reverse bias, $C(V) < C(0)$ and the imaginary part is negative. For forward bias, $C(V) > C(0)$ and the imaginary part is positive. It should be noted this normalized, matched varactor plot also corresponds to a plot of net Q-values as Equation 3.47a can be rewritten as follows with $r_s = 1$.

Large-Signal Characteristics

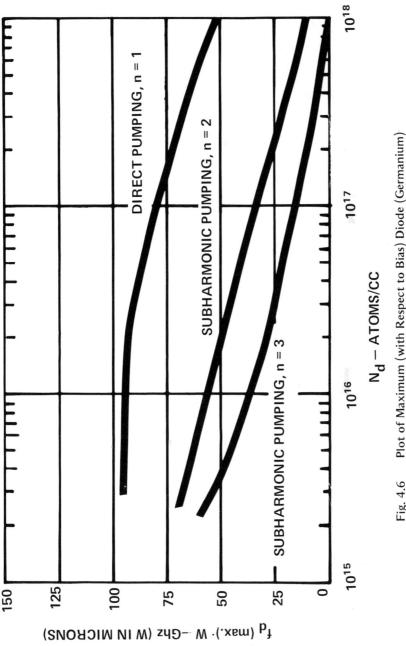

Fig. 4.6 Plot of Maximum (with Respect to Bias) Diode (Germanium) Figure of Merit as a Function of Base Doping

indicate the significance of a pump power constraint on the variation of $f_d(max)$ with doping, the pump voltage required to produce f_d (max) for a base doping of $3 \cdot 10^{17}$ atoms/cc was selected and used to determine f_d for the lighter base dopings with Figure 4.5. The dash curve drawn in Figure 4.6 indicates the sharply reduced figure of merit obtainable under such pump power constraints. Thus if low pump power is mandated, lower breakdown voltage units with higher base doping represent the better performance choice.

It is interesting at this point to compare f_d (max) to the earlier figure of merit, namely, the diode cutoff frequency, $f_c(V_b)$. The comparison of $f_d(max)$ with $f_c(V_b)$ for the abrupt junction can be readily gained by recognizing that, for $f_d(max)$, C_0 is approximately specified at $V_b/2$ or is equal to $\sqrt{2}C_{min}$. Making this substitution into Equation 4.11 with $b_0 \approx 1$, $f_d(max) \approx 0.354\ a_n/a_0^2 f_c(V_b)$ (for abrupt junction), and in a similar manner $f_d(max) \approx 0.397\ a_n/a_0^2 f_c(V_b)$ (for graded junction). For a doping such that $\gamma_{max} = 0.9$, $a_1/a_0^2 \approx 0.50$ (abrupt junction), $a_1/a_0^2 \approx 0.38$ (guided junction), $f_d(max) = 0.18\ f_c(V_b)$ for the abrupt junction and $0.15\ f_c(V_b)$ for the graded junction. At best, the maximum ratio of a_1/a_0^2 for the abrupt junction using sinusoidal pumping is 0.557 (occurring for $\gamma = 0.992$) such that the greatest value of $f_d(max)$ is $0.20\ f_c(V_b)$ (abrupt junction, m = 1/2). The greatest value for the graded junction (m = 1/3) is approximately $0.17\ f_c(V_b)$. Thus, it can be concluded that $f_d(max)$ can range up to 20% of the diode cutoff frequency depending on the maximum relative permissible swing and the diode type.

To illustrate the use of the diode figure of merit, f_d, and present diode characteristics with their potential amplifier performance, the following varactor and amplifier situations are considered. With present day techniques and materials, it is possible to obtain gallium arsenide varactors of the Schottky barrier type (abrupt junction characteristic) which posses $f_c(0) \approx 300$ GHz and $C(0) \approx 0.02$ pF. Such varactors if biased for amplifier operation at $V_0 = -2V$ would have $C_0 = 0.01$ pF, $f_{co} = 600$ GHz, and $R_s = 27$ ohms. If it is assumed that $\phi_0 \approx 0.8V$ and $V_{fwd} = 0.5V$ without significant shot noise being introduced, then the maximum useful relative pump swing, γ, is about 0.89 yielding a value of $a_1/a_0^2 \approx 0.50$ from Figure 4.3. From Equation 4.11c, if b_0 is assumed to be unity, it is evident that the diode figure of merit in this operating condition is $f_{d0} = 150$ GHz. Such a varactor operating figure of merit can yield attractive amplifier performance even at millimeter wavelengths. For examples, consider an amplifier operating at 34 and 92 GHz (8 and 2 mm). For maximum performance (minimum noise figure), the idle frequency would be chosen equal to the operating point diode figure of merit, i.e., $f_2 = f_{do} = 150$ GHz. With the above signal and idle frequencies specified, the required pump power would be supplied at 184 GHz and 242 GHz respectively. Under such operating conditions, the amplifier whose signal

Large-Signal Characteristics 75

frequency was 34 GHz' could possess a minimum noise figure of 1.6 dB (T_a = 135°K) as determined from Equation 4.13. Similarly, a 3.5 dB (T_a = 370°K) performance would be obtained at 92 GHz. The pump power dissipation in the varactor for this operation ($V_m = V_0 + V_{fwd}$ = 2.0 + 0.5 = 2.5V) would be 21 and 35 mW respectively for the two required pump frequencies. In actual amplifier configurations at these frequencies, however, considerably greater pump power sources would most likely be required to supply the circuit transmission and matching losses in addition to the power dissipated in the varactor losses. Nevertheless, the calculated amplifier performances cited are typical of what could be obtained as determined by the time-dependent element evaluation presented here.

4.1.4 Limitations on Time-Dependent Evaluation

When employing the time-dependent element model for the large-signal representation of the varactor, several factors regarding the applicability of this characterization must be noted. First, the simple model of the varactor element as a frequency-invariant element is not always valid. It is assumed, for example, that the voltage-dependent capacitance of the diode can be converted to a time-dependent capacitance by simply determining the Fourier coefficients of the capacitance waveform generated by the application of a pump voltage. If, however, the C-V law is frequency-dependent (time rate of change dependent), then a separate evaluation of the coefficients at each frequency of concern must be made. Such an evaluation results in a different diode figure of merit for each frequency of concern also. This latter fact can be accounted for in the noise figure expression (Equation 4.10), however, by replacing the single value of f_d^2 by the product of the values at the signal and idle frequencies, i.e., by $f_{d_1} f_{d_2}$. Generally, the C-V law is frequency-invariant over its useful frequency range except for charge storage effects (injection capacitance) with forward conduction or because of geometry and skin effects resulting in transverse wave propagation within the junction region. As shot noise must generally be kept to a minimum in amplifier or converter applications where the time-dependent element approach is most useful, little problem is encountered with storage effects if they exist (Note: Schottky barrier units and short lifetime materials, e.g., GaAs, have none). Furthermore, the second factor determining the frequency invariancy of the C-V law is generally negligible as the junction diameters employed are seldom even a significant fraction of the wavelengths involved. Thus, in general, the evaluation of the time-dependent capacitance coefficients by the methods used from the C-V law is adequately justified for low noise parametric amplifier or converter operation. Perhaps the element to be found most frequency-dependent, if any are, particularly at the higher frequencies, is the varactor series resistance which is at least partially controlled by skin effects ($f^{1/2}$ dependence).

However, as this resistance is generally not appreciably voltage-dependent, no time-dependent evaluation is required and any frequency dependence can be accounted for in the figure of merit values as described above.

The second point to be made regarding the applicability of the time-dependent evaluation made here is the influence of the diode or varactor temperature behavior. Specifically, no account has been made of the pump power dissipation within the varactor which, of course, can cause a substantial rise in device temperature or of significantly different environmental or ambient operating conditions. Either of these conditions can invoke operation of the device at temperatures both well above and well below room temperature. Thus it must be recognized, particularly for large-signal operation, that the device parameter evaluations as made in this section must be carefully re-evaluated for the operating temperature intended or reached. This is particularly important in the use of f_d for not only are both C_0 and R_s temperature-dependent as discussed in the previous chapter but the temperature dependence of ϕ_0 can markedly affect the relative pump swing, γ, and the capacitance and resistance coefficients, a_n and b_n. All of these factors can be duly accounted for, however, by appropriately establishing the device operating temperature and employing the proper device values. Establishment of the operating temperature if dissipation is significant can be gained by suitable analysis as discussed in Chapter VI.

The third and final point regarding the limitations of this varactor characterization approach or evaluation is the assumption of a sinusoidal pump voltage being applied to the junction. This assumption, which implies a voltage source as seen by the varactor (nonlinear and/or voltage-dependent element), permits the direct analytical evaluation of the Fourier capacitance and resistance coefficients. If impedances exist in the pump source-varactor loop at harmonics of the pump frequency, then voltage sources at these frequencies can be developed by the harmonic currents produced by the sinusoidally driven nonlinear capacitance. In this manner, by superposition, the applied varactor pump voltage may not be sinusoidal and the coefficients would differ from those evaluated here. It might be noted, however, that other waveforms such as a square wave could yield higher values of f_d for the same varactor provided such waveforms could be impressed. Nevertheless, because of the difficulty in coefficient evaluation or re-evaluation, the time-dependent element approach presented here is generally of greatest value in the design of low-noise (small-signal) parametric amplifiers or converters which can be designated as large-signal—small-signal design problems. In the event that harmonic generation is the goal, then a large-signal—large-signal design analysis is required. Such a design can be carried out using the basic approach presented here (assuming the injection or diffusion capacitance is also accounted for) but requires reiterative, numerical evaluation of the coefficients for a nonsinusoidal pump (junction) voltage. A less precise (accounting for gross changes only) but

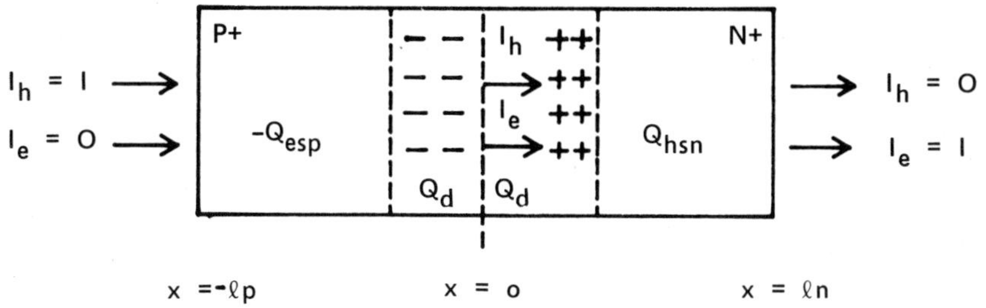

Large-Signal Characteristics 77

```
                P+      | - -  | I_h ++|      N+
I_h = 1  →              | - -  |  →  ++|           →  I_h = 0
I_e = 0  →     -Q_esp   | - -  | I_e ++|  Q_hsn    →  I_e = 1
                        | - -  |  →  ++|
                        |  Q_d | Q_d   |

        x = -ℓ_p                x = 0                x = ℓ_n
```

Fig. 4.7   Varactor Charge Distribution

more realistic approach and model for this latter case including injected charge is presented in the next section.

4.2   Varactor as a Charge Storage Element

4.2.1 Charge Control Equations

The analysis and particularly the evaluation presented in the previous section was based upon the assumption that the voltage variation across the varactor was sinusoidal and such as to produce negligible effects due to injection or diffusion capacitance. If the voltage swing is such as to bias the varactor in the forward conduction region, a more fruitful approach for determining the varactor performance is to consider the varactor as a charge storage tank whose charge depends upon the current flow. The analysis is based upon the charge control equation that is now described. Figure 4.7 is a schematic representation of the charges and currents in a typical diffused varactor. The point $x = 0$ represents the metallurgical junction and the end points $x = \ell_n$ and $x = -\ell_p$ represent, respectively, the ohmic contacts at the ends of the N^+ and P^+ regions. The charges Q_{d+} and $-Q_{d-}$ represent the charge of the ionized donors and acceptors in the depletion region. The charge Q_{hsn} represents the positive charge due to holes stored in the quasi-neutral N region. Because of the requirement of charge neutrality, an equal and opposite number of electrons is stored in the same region, but it is not indicated in the diagram. In the same way, $-Q_{esp}$ represents the negative charge due to electrons stored in the quasi-neutral P region. Again, charge neutrality requires an equal and opposite number of holes stored in the same region. With the assumption of highly extrinsic P^+ and N^+ regions, the minority carrier currents at the end of these regions are zero, that is $I_e(x = -\ell_p) = 0$ and $I_h(x = \ell_n) = 0$. Noting the above definitions and conditions, the equations which express particle and charge conservation can now be written. The relations which express the continuity of the hole and electron currents are

$$I_h(x=0) = \frac{dQ_{hsn}}{dt} + \frac{Q_{hsn}}{\tau_h} \qquad 4.16$$

$$I_e(x=0) = \frac{dQ_{esp}}{dt} + \frac{Q_{esp}}{\tau_e} \qquad 4.17$$

where τ_h and τ_e are respectively the lifetime for the minority carriers in the N and P regions. The equations which indicate conservation of charge are

$$-I_e(x=\ell_n) + I_e(x=0) + I_h(x=0) = \frac{dQ_{d+}}{dt} \quad \text{(right-hand side of diode)} \qquad 4.18a$$

$$\text{or} \quad -I_e(x=0) + I_h(x=-\ell_p) - I_h(x=0) = -\frac{dQ_{d-}}{dt} \quad \text{(left-hand side of diode)} \qquad 4.18b$$

with $I_e(x=\ell_n) = I_h(x=-\ell_p) = I$ \qquad 4.19

where I is the current at the terminals of the device. These last equations neglect any displacement current which may be present in the quasi-neutral N and P regions. Combining the above Equations (4.16, 4.17, 4.18a and 4.19), the following current expression is obtained

$$I = -\frac{dQ_{d+}}{dt} + \frac{dQ_{hsn}}{dt} + \frac{dQ_{esp}}{dt} + \frac{Q_{hsn}}{\tau_h} + \frac{Q_{esp}}{\tau_e} \qquad 4.20$$

This equation is the charge control equation and indicates that the terminal current supplies the charge necessary for recombination and for storage in the quasi-neutral regions and in the depletion layer. There are two typical cases that simplify the form of the above equation. First, the situation in which appreciable charge injection takes place such that the current contributed by the depletion layer charge, Q_d, is negligible and, second, when the reverse situation exists and no charge is injected or stored other than that of the depletion region. The charge injection case, applicable particularly to the graded junction silicon varactor, is of the greatest interest because of the substantial charge exchange and correspondingly effective harmonic generation. The no-injection or short-lifetime case, germane to the Schottky barrier type diodes or units fabricated from gallium arsenide is concerned only with the voltage-dependent depletion layer charge or capacitance even with forward bias operation. This case is therefore best handled by the reiterative, time-dependent capacitance or charge approach mentioned in the previous section. The balance of this section therefore treats the charge control equation and circuit model for the important charge injection dominated case.

Large-Signal Characteristics

Considering the injected charge to dominate the current, two simplifications to the charge control equation (4.20) can be made. First,

$$\frac{dQ_{d+}}{dt} \ll \frac{d}{dt}(Q_{hsn} + Q_{esp}) \qquad 4.21$$

Second, if the carrier storage and removal in the quasi-neutral region takes place during a time smaller than the minority lifetimes τ_h and τ_e, then the carriers lost by recombination are effectively zero, that is $Q_{hsn}/\tau_h \to 0$ and $Q_{esp}/\tau_e \to 0$. With these two approximations the charge control equation reduces to

$$I = \frac{dQ_s}{dt} \qquad 4.22$$

where Q_s represents the stored charge in the quasi-neutral regions. The charge control equation, by itself, is not enough to determine the circuit operation of a varactor. It is necessary to have a circuit model for the varactor which, together with the charge control equation, are sufficient to determine or specify the operation of the device. This subject is discussed next.

4.2.2 Circuit Model

When employing the charge injection mode of operation just described in a harmonic generator design, it is necessary to identify and combine the V-I characteristic of the device with the charge control equation to obtain a complete circuit model. Figure 4.8 shows the static V-I characteristics of a P-N diode as well as the characteristics of an ideal diode which has zero voltage drop in the forward direction. An approximate model for the dynamic V-I characteristics (injected charge dominated) at high frequencies ($1/2f \ll$ transit or lifetime of injected carriers) can be obtained by coupling the static V-I characteristics with the charge control equation. This is done as follows. Figure 4.8 shows an ac voltage applied across the varactor. During the time that the voltage drives the varactor in the forward direction, the current is determined by either the non-ideal V-I characteristic or by the external circuit in the case of the ideal characteristic. When the voltage becomes negative, the diode still remains in a low impedance condition because of the stored charge and remains so until the reverse current removes the stored charge. From there on, the varactor changes to a high impedance state and the current is essentially zero. The performance of the varactor as a harmonic generator can be obtained by a Fourier analysis of the current and the voltage waveforms in order to determine the power content of the different harmonics. This Fourier analysis can be carried out either by analytical means or by numerical methods using a digital computer.

A current waveform with a high content of harmonics is obtained if, after the removal of the stored charge, the transition from the low impedance state to the

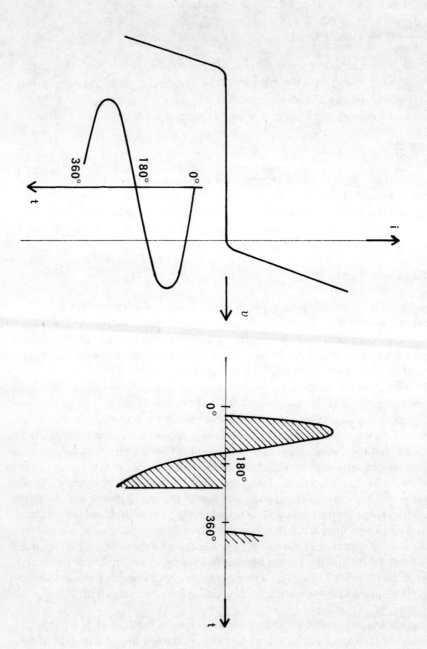

Fig. 4.8 Large-Signal, Varactor Circuit Model

Large-Signal Characteristics

high impedance state is achieved in as short a time as possible. Varactors emphasizing this characteristic have been designed and are known as "snap diodes." The effect is accomplished by suitably grading the base region to create a retarding potential enhancing the charge stored near the junction (see discussion on injection capacitance, Section 2.2.1 and 3.2.3).

In this model of the varactor and in the ensuing analysis, it is assumed that after the removal of the charge stored the current flowing through the diode is small. As the complete charge equation indicates, there is current flow caused by the change in the charge in the depletion layer. If the time rate of change of the applied voltage is high enough, this current may not be negligible and, as a matter of fact, imposes an upper frequency limit to this charge injection mode of operation of the varactor as a harmonic generator. Beyond this mode of operation, of course, the varactor can still technically function as a harmonic generator element by virtue of its voltage-dependent capacitance or charge as previously noted. In actual fact, however, this transfer of dominance from injected to depletion layer charge control with frequency occurs at or near the upper useful limit of the varactor as determined by the basic charge relaxation process. This latter limit can be estimated from the small-signal, zero-voltage cutoff frequency value, $f_c(0)$. The treatment of the varactor as a charge storage element presented in this section is a very simplified treatment. More detailed discussions of the operation and equivalent circuit of snap diodes can be found in the literature.[4]

4.3 Varactor or PIN as a Switching Element

4.3.1 Device Design and its Two-States

The varactor can also be employed as a two-state switching element as suggested in Section 1.2 and depicted in Figure 1.1b. The two impedance states are obtained through changes in bias level, generally from forward to reverse, with substantial change in the device capacitance. Such an element can be made to provide the two impedance states desired for RF and microwave switching applications provided the signal levels are kept small compared to the bias voltages applied. Attempts to operate the conventional varactor as a switching element for large signals results in significant changes in the RF waveform because of the voltage-dependent capacitance and/or rectification character of the device as well as causing shifts from the bias defined, small signal impedance states. To permit satisfactory switching of large signals as well as to increase the ratio of the two-state impedance levels for maximum insertion loss change as a switching element, a punch-thru varactor (depletion layer extending through the base region) is used. The extreme case of this nonvariable capacitance device which is specifically designed for switching purposes is the PIN diode. The PIN is a P-N junction device

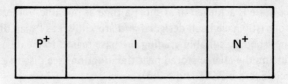

(A) PHYSICAL STRUCTURE

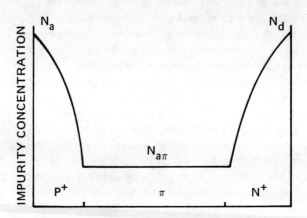

(B) DOPING PROFILE OF A DIFFUSED $P^+\pi N^+$ UNIT

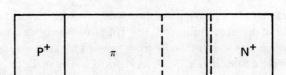

(C) DEPLETION LAYER IN A $P^+\pi N^+$ UNIT

Fig. 4.9 The PIN Diode — (a) Physical Structure, (b) Doping Profile of a Diffused $P^+ \pi N^+$ Unit, (c) Depletion Layer in a $P^+\pi N^+$ Unit

Large-Signal Characteristics

like the conventional varactor but with a doping profile such that an intrinsic layer, the "I-region," rather than an N base region, is sandwiched in between a P^+-layer and an N^+-layer as shown in Figure 4.9a. In a practical device, the I-region is approximated by either a high resistivity P-layer (π-layer) or a high resistivity N-layer (ν-layer). The doping profile of a diffused $P^+ \pi N^+$ structure is shown in Figure 4.9b.

Under no-bias condition, a depletion layer is formed at the junction of the N^+- and π-layers as shown in Figure 4.9c. The extent of this depletion layer is a function of the resistivity of the π-layer. With a reverse bias applied to the diode (P^+-region negative), mobile carriers are removed from the π-layer and the depletion layer penetrates further into the π-region. If sufficient reverse bias voltage is applied to the diode, it is possible to remove all the mobile carriers from the π-layer and the depletion layer will extend throughout the π-region. This state of the depletion layer, as noted earlier, is called the punch-thru condition. Any further increase in voltage will not increase the thickness of the depletion layer by any appreciable amount, because of the high doping in the P^+-region; however, the electric field in the π-region always increases with applied reverse voltage. If the electric field in the depletion layer is high enough, avalanche breakdown occurs. Which of the two conditions occurs first, punch-thru or avalanche breakdown, is determined by the thickness and resistivity of the π-region. Generally, PIN's are fabricated such that punch-thru occurs at one-tenth or less of the breakdown voltage.

With a forward bias voltage applied across the diode, the P^+-region positive with respect to the N^+-region, holes from the P^+-region and electrons from the N^+-region are injected into the π-region. The distribution of carriers is shown in Figure 4.10. The injected carriers create a quasi-neutral plasma which increases the conductivity of the π-region typically by as much as a factor of 10^2 to 10^5. The number of carriers injected into the π-region depends upon the forward bias current, upon the thickness of the π-region and upon the lifetime of the injected carriers. Another effect of the applied voltage is to decrease the width of the depletion layer between the π- and N^+-regions. Benda and Spenke[5] give a detailed discussion of the forward biased state and of the switching process from the forward into the reverse state of PIN diodes.

Because of the transit time considerations, a PIN diode becomes a poor rectifier at frequencies higher than a few megahertz. That is, the excess carrier distributions and depletion layers created by a dc bias signal are not affected by superimposing upon the dc bias an RF signal of high enough frequency. Thus at RF and higher frequencies, the dc bias sets the impedance state of the device virtually independent of the relative amplitude of any RF signal applied. It is therefore

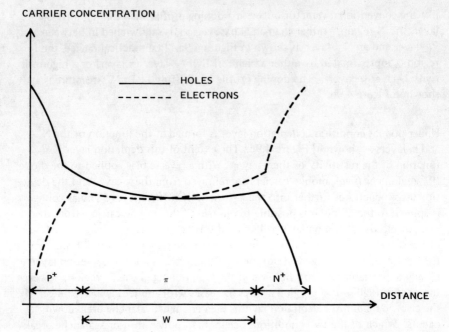

Fig. 4.10 Schematic Representation of the Hole and Electron Distribution in a Forward Biased $P^+\pi N^+$ Diode

possible to control RF and microwave current amplitudes with forward bias which are orders of magnitude greater than the bias current and voltage amplitudes with reverse bias which can be nearly equal to the breakdown voltage. The RF control gain (ratio of RF to dc bias power) is thus quite substantial, ranging from 10^3 to 10^4 for well fabricated PIN microwave diodes.

4.3.2 RF Equivalent Circuits

Assuming RF operation of the device as described above, four regions can be distinguished in a $P^+ \pi N^+$ diode. Two of these regions are the highly doped P^+- and N^+-regions where characteristics are in the first approximation, independent of the bias applied. The other two regions are the depletion layer formed at the junction of the N^+- and π-regions and the π-region itself. Therefore, the RF equivalent circuit of the $P^+ \pi N^+$ diode is as shown in Figure 4.11a. The resistances R_{ps} and R_{ns} represent the ohmic resistance of the P^+- and N^+-regions. For abrupt junctions, these resistances are independent of bias and their value may change with frequency because of the skin effect. In the punch-thru condition,

Large-Signal Characteristics

which is a desirable mode of operation in microwave switching applications, the undepleted π-region is absent and the conductance G_j of the depletion layer is small. Therefore, the equivalent circuit reduces to the one shown in Figure 4.11b, which can also be represented as a parallel combination of a capacitance C'_j and a resistance R_{rp}. If the Q of the diode is high enough, C_j and C'_j are approximately the same. Olson[6] presents the derivation of design formulas for predicting the RF impedance of diffused PIN diodes as a function of reverse bias voltage. Under forward biased conditions, the capacitance and conductance of the depletion layer become very large and can be neglected. Furthermore, since the resistance of the π-region is decreased by carrier injection (conductivity modulation), the capacitance C_π can also be neglected. The resulting equivalent circuit is shown in Figure 4.11c. The three resistances can be lumped in a single, equivalent forward series resistance R_{fs}. The contribution of R_{ps} and R_{ns} to the total series resistance R_{fs} depends upon the thickness of the P^+- and N^+-regions and the thickness of and injection level into the π-region. In well fabricated $P^+ \pi N^+$ diodes, R_π is almost equal to R_{fs} except at the highest bias currents. The calculation of R_π, that is the RF resistance of the π-region under forward bias condition, has been calculated by Leenov[7] for the case of equal mobilities of electrons and holes. The value of R_π for the case of unequal mobilities has not been given as such in the literature but can be obtained from expressions presented there.[5] The resultant expression obtained for the RF resistance, dc bias current product is as follows:

$$R_\pi I_{dc} = \frac{kT}{e} \frac{8b}{(1+b)^2} \frac{\sinh W/2L}{\sqrt{1-B^2\tanh^2 W/2L}} \tan^{-1}[\sqrt{1-B^2\tanh^2 W/2L} \sinh W/2L] \text{ in volts}$$

4.23

where I_{dc} is the forward bias current resulting in the RF, π-region resistance, R_π, W is the π-region width, L the ambipolar carrier diffusion length, b the mobility ratio, $B = (b-1)/(b+1)$, k the Boltzmann constant, e the electronic charge and T the temperature in °K. The plot of the $R_\pi I_{dc}$ product so obtained, where I_{dc} is the forward bias current, taking into account the difference in hole and electron mobilities in silicon is given in Figure 4.12.

The equivalent circuits given in Figures 4.11b and c are exclusive of the parasitic elements introduced by the diode package and mount. These parasitic elements become important and are usually dominant at X-band and higher frequencies. (See Section 5.1.1.) A general transformation representation for a packaged diode in a transmission line media is given in Section 5.1.2 which allows the diode package and mount to be represented by a three-port transformation matrix (or equivalent circuit), one port of which is connected to the diode element terminals and the other two to the input and output of the transmission line system.

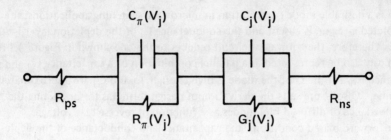

(A) COMPLETE EQUIVALENT CIRCUIT

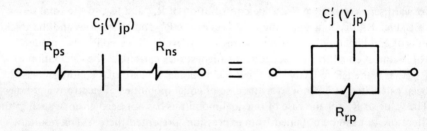

(B) EQUIVALENT CIRCUIT UNDER PUNCH-THRU CONDITIONS

(C) EQUIVALENT CIRCUIT UNDER FORWARD BIAS

Fig. 4.11 The RF Equivalent Circuit of the PIN Diode — (a) Complete Equivalent, (b) Equivalent Circuit Under Punch-thru Conditions, (c) Equivalent Circuit Under Forward Bias

Large-Signal Characteristics

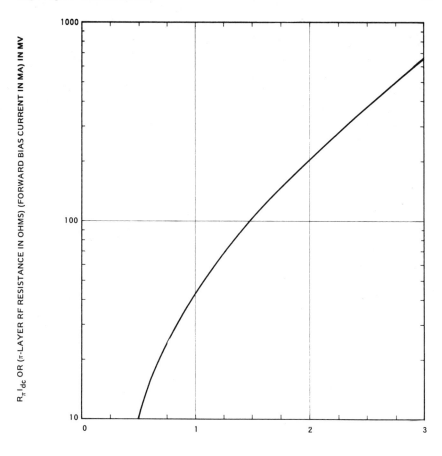

W/2L (π-LAYER THICKNESS/2 · DIFFUSION LENGTH)

Fig. 4.12 Plot of the $R_\pi I_{dc}$ Product of a $P^+ \pi N^+$ Structure as a Function of $\frac{W}{2L}$ (T = 300°k)

4.3.3 Other Characteristics

In the application of PIN diodes to microwave switching, the transition period from one bias state to the other or switching speed is an important performance parameter. The switching speed of the PIN device is basically defined by the time required to create or remove the I-region plasma in obtaining the high conductance or low susceptance state. As the time required for injection or removal of carriers to an equilibrium state depends on the distance they must travel (on the average, one-half the I-region width, W, assuming parallel plane junctions and no charge storage outside the junction area) and their respective velocities (approaching about 1% of the carrier saturation velocities, equal to about 10^7 cm/sec. in silicon), the switching speed or time is of the order of $50W/v_s$. Thus for a silicon PIN diode with an I-region width of about 30μ ($V_b \approx 500$ V), the switching speed is of the order of 15 ns. In most instances, some charge is stored beyond the volume covered by the junction areas and the electric fields created in the I-region are not sufficient to cause the carriers to achieve as much as 1% of their saturation velocities resulting in somewhat longer switching times ranging up to 100 ns. Furthermore, because the saturation velocities are not generally achieved, the velocities are field-dependent. As the average electric field sustained in the depleted portion of the I-region during the turn-on time (plasma buildup from injection) is generally higher than that achieved during the turn-off time (plasma extraction) because of the difference in the establishment of the two states, the turn-on time is usually less than the turn-off time for PIN's. A complete discussion of the reverse recovery process is given by Benda and Spenke[5]. No equivalent discussion of the turn-on process with its associated transition period is yet available as the nonlinear nature of an accurate analysis makes anything but a specific numerical analysis very difficult.

In addition to the transition period, the RF impedance exhibited by the diode during the transition between states is also important. As the diode plasma is created or removed in the switching of states, the impedance can be largely resistive (depending upon the I-region capacitance and frequency of operation) and of such values as to pass close to or through an impedance match condition of any transmission media in which it is in shunt (e.g., as in a simple transmission line switch configuration). Under such operating conditions, the diode can absorb and dissipate for a portion of the transition period up to approximately one-half of the available transmission line power in sharp contrast to the usual 100/1 power mismatch in the two bias states. Thus, with continuous RF power applied during switching, the impulse of dissipation which can occur with its corresponding device temperature rise must be recognized and minimized in component design. (See Section 6.2.) An experimental evaluation of the particular PIN transition impedance[8] is therefore often an important characteristic to obtain.

In determining the two-state limitations of the PIN diode in RF or microwave switching and phase shifting applications, Hines[9] has shown that the power handling capability of these microwave components is directly proportional to the product of the maximum RF voltage the diode can withstand in the reverse

Large-Signal Characteristics

bias condition and the RF current it can safely carry in the forward bias state, independent of the two-port, lossless network used for interconnecting the PIN diode element and the transmission line. Furthermore, he defines a two-state cutoff frequency for PIN diodes used in microwave switching and phase shifting which determines the loss ratio (ratio of power dissipated to incident power) in both the transmit and reflect states. This switching cutoff frequency, like that for the varactor (as a single state capacitor), is a useful diode figure of merit and is defined as follows:

$$f_{cs}(\text{switching}) = \frac{1}{2\pi C_j \sqrt{R_{fs} R_{rs}}} \text{ in c/s} \qquad 4.24$$

where R_{fs} is the total forward bias state series resistance, R_{rs} is the reverse bias value and C_j is the reverse bias state junction capacitance as defined in the previous section. In practice, additional mounting and package capacitance may be added to C_j to obtain a more representative working value for f_{cs}. (See Section 5.1).

The limitations imposed on use of the varactor or PIN by the dissipation occurring in either operating bias state is generally determined by the device operating temperature restrictions. The temperature rise of the diode under pulse power dissipation for both bias states is presented in Section 6.2.

REFERENCES

1. K.E. Mortenson, "Parametric Diode Figure of Merit and Optimization," *J. of App. Phys., 31*, 1207-1212 (July 1960).

2. D.B. Leeson, "Capacitance and Charge Coefficients for Varactor Diodes," *Proc. IRE, 50* (August 1962).

3. P. Penfield and R.P. Rafuse, "Varactor Applications," M.I.T. Press, Cambridge, Mass. (1962).

4. J.L. Moll and S.A. Hamilton, "Physical Modeling of the Step-Recovery Diode for Pulse and Harmonic Generation Circuits," *Proc. IEEE, 69*, 1250-1259 (July 1969).

5. H. Benda and E. Spenke, "Reverse Recovery Processes in Silicon Power Rectifiers,"*Proc. IEEE, 55*, 1331-1354 (August 1967).

6. H.M. Olson, "Design Calculations of Reverse Bias Characteristics for Microwave PIN Diodes," *IEEE, ED-14*, 418-428 (August 1967).

7. D. Leenov, "The Silicon PIN Diode as a Microwave Radar Protector at Megawatt Levels," *IEEE, ED-11*, 53-61 (February 1964).

8. R. Galvin, and A. Uhlir, Jr., "Transient Microwave Impedance of p-i-n Switching Diode," *IEEE, ED-11*, 441 (1964).

9. M.E. Hines, "Fundamental Limitations in RF Switching and Phase Shifting Using Semiconductor Diodes," *Proc. IEEE, 52*, 697-708 (1964).

BIBLIOGRAPHY

Kotzebue, K.L. "Circuit Model of the Step-Recovery Diode," *Proc. IEEE, 53*, 2119-2120 (December 1965).

Moll, J.L., Krakauer S. and Shen, R., "P-N Junction Charge Storage Diodes," *Proc. IRE, 50*, 45-53 (January 1962).

Penfield, P. Jr., "Fourier Coefficients of Power-Law Devices," J. Franklin Inst., *273*, 107-122 (February 1962).

Scanlan, J.O. "Analysis of Varactor Harmonic Generators," *Advances in Microwaves, 2*, ed. L. Young, Academic Press, Inc., N.Y., N.Y. (1967).

PACKAGING CONSIDERATIONS 5

5.0 General

As the package which houses or encapsulates the variable capacitance diode has a pronounced influence on the use of such elements, it is pertinent to discuss here in some detail several of the features and problems resulting from packaging. Specifically, there are three major properties which are of interest to the user of such devices; namely, electrical, thermal and mechanical. Chief attention will be given here to the electrical properties of the packaging as these are of primary importance in device utilization and component circuit design. In addition, some discussion is also pertinent regarding the thermal properties of such packaging as these will determine the power handling capability of these units under various circumstances. Finally, the mechanical properties of conventional packages are so varied that they are generally specified in detail by the manufacturer; consequently no attempt to discuss these properties will be made here.

5.1 Electrical Properties

5.1.1 Parasitic Elements

Before discussing the electrical influences that the package may exert on the appearance to the outside world of the variable capacitor element contained within, it is important to recognize what package or parasitic circuit elements can be physically contributed by the package or encapsulation. It would be impossible here to consider all the variations in package designs which are available on the market and which can differ markedly depending upon what frequency range and use for which they are intended, but, as most of these

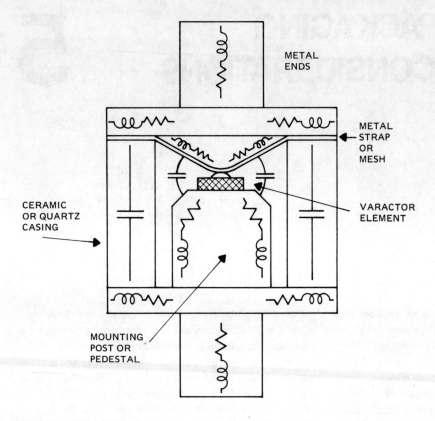

Fig. 5.1 Typical Variable Capacitance Diode Package Cross-Section and the Parasitic Elements

packaged structures do contain many of the same parasitic elements, it is useful to consider a typical varactor package structure. Such a structure is shown in Figure 5.1. In this case, it is seen that the package consists of an insulating spacer or casing separating two metallic end pieces sealed in such a manner as to provide a hermetic encapsulation for the semiconductor element within. Within the package, a semiconductor element is usually mounted on a post or pedestal as shown with a suitable strap or mesh arrangement making contact to the opposite end of the diode element. As depicted, both metallic and ceramic parts contribute inductance and capacitance between the contacts to the actual semiconductor element and the connections which are made externally to the diode housing or package.

Considering first the capacitive elements of the package, appearing in shunt with

Packaging Considerations 93

the variable capacitive element (although possibly separated by inductance) exists the capacitance between the upper contact and the main body of the semiconductor element and its metallic mount. Because of the close spacing required, particularly for small junction elements which also possess very small junction capacitances, this capacitance contribution can be quite significant. Such is typically the case, for example, for the very high frequency microwave and millimeter wave diodes used. It should also be noted here that any significant dielectric encapsulation in direct contact with the semiconductor element still further enhances this parasitic contribution and must be recognized in spite of the other advantages of such encapsulation. In addition to the contact and strap capacitance bridging across the semiconductor element, there also exists a generally much larger capacitance made up of the insulating casing or housing. This capacitance is generally an order of magnitude larger than the capacitance just described and is therefore quite significant although possibly more removed from the diode element itself. Obviously it behooves the diode fabricator to use as low a dielectric constant material and appropriate casing geometry consistent with mechanical strength requirements to minimize this parasitic contribution. For this reason, although ceramic casings are generally employed, in some instances quartz, with less than half the dielectric constant of the usual ceramics, is employed. Further, with either dielectric an effort is made to provide the longest possible path for displacement currents commensurate with maintaining the minimum inductive contributions to the package.

In addition to the capacitances, all the metallic portions of the package of course contribute inductance which appears in series with the variable capacitance element as well as the various parasitic capacitances. Generally speaking, the most significant contributions to this inductance come from the metallic parts within the casing, namely the contacting strap or mesh and the pedestal or post upon which the semiconductor element is mounted. These contributions are most significant because of the very small cross-sectional dimensions of these parts with lengths which are comparable to the dimensions of the package. For units which are intended for high frequency applications, particular attention is given to the design of these inner parts particularly the strap connection, to minimize this inductive contribution; thus, where lengthy internal connecting leads may be tolerated in units for use at the lower RF frequencies, very short strap, mesh or diaphragm configurations are utilized in the high frequency units.

Associated with both the capacitive and inductive parasitic elements are losses which can deteriorate the apparent quality of the variable capacitive element. For most frequencies of interest, with the possible exception of the very high millimeter wave region, the insulator materials can generally be considered lossless; however, as the current flow in the metallic parts of the package are generally skin effect limited over virtually the entire useful range of this type of device, these parts can contribute losses significant when compared to the losses contained within the semiconductor element itself and therefore must be accounted for when using these elements. To meet the thermal expansion matching requirements, as well as for metallizing purposes, the metallic parts of these package structures are generally not made of good conductors and are therefore plated (generally gold). At the metallizing point which joins the casing to the end pieces, however, no plating exists and indeed relatively poor conductive materials are involved which can be a principal source of package losses as the currents must flow along the surfaces of the conductor. Another primary contributor to the losses of the package comes from the contact to the end pieces of the package. These contacts, which for replacement purposes are generally not fixed but of a pressure variety, can also significantly add to the total losses and are generally attributed to the package.

5.1.2 Package Equivalent Circuits

First Order of Approximation: At sufficiently low frequencies, parasitic package elements can be lumped together to give a relatively simple package, equivalent circuit. Such a circuit is illustrated in Figure 5.2 where it is assumed that all of the series inductance contributions between the contacts to the diode and the outside world can be lumped into one series element, L_p, and all of the shunt capacitance contributions can be lumped together into a single shunt element, C_p. In this simple approximation, it is generally most useful to consider the inductance of the package as being shunted by the total package capacitance which is considered to consist primarily of the casing capacitance as at the frequencies where this circuit is most appropriate, the junction capacitance is much more significant than any localized parasitic capacitance which may shunt it. Typical values for the series inductance of such an equivalent circuit range from a few tenths to several nanohenries where the package capacitance may be as small as a few hundredths of a picofarad for quartz encased units up to several picofarads for the larger ceramic packages utilized at the lower frequencies. Although it is difficult to make any general statement regarding the applicability of this simple package equivalent circuit, it is typically useful for junction capacitances ranging from the highest values in the order of tens of picofarads down to the order of one picofarad and for frequencies ranging up into the UHF region. Package losses under these operating circumstances can generally be neglected.

In utilizing this simple two-element, equivalent circuit, it is appropriate to point out that under certain operating conditions this can be simplified still further. For example, at RF and, in many cases, even up into the VHF range, the inductive reactance associated with L_p is small enough to be neglected in comparison

Packaging Considerations

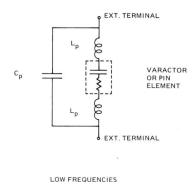

Fig.5.2 Lumped, Equivalent Package Circuits (Neglecting Losses)

with the much larger reactance of the diode element itself such that the equivalent package circuit becomes simply a capacitance, C_p, in shunt with the variable capacitance provided by the diode junction. Such capacitance has the principal effect of diluting the percentage change in capacitance obtainable with the variable capacitive element as in many instances it can be simply tuned out along with the junction capacitance at a selected bias point. For another example, there are instances at the higher end of the frequency range where this simple two-element, equivalent circuit is appropriate where the package capacitance can be ignored and only the series inductance, L_p, considered. Such is the case when the variable capacitive element is driven sufficiently hard or possibly biased into the forward conduction region such as is appropriate for limiting or switching use that the junction impedance becomes quite low, reaching values comparable to and even less than the inductive reactance.

A More Exact Lumped Model: Diodes employed in the UHF and microwave region can often be better represented by a more distributed four-element equivalent circuit forming a filter structure. Such a representation is also depicted in Figure 5.2 wherein the capacitance and inductive elements have each been split into two component parts such that the capacitance in the immediate vicinity of the semiconductor element forms the initial shunt element, C_{p2}, with the strap and post or pedestal inductances appearing immediately in series with the semiconductor element past the initial shunt capacitance as represented by L_{p2}. Moving out to the terminals of the package still further, there next appears the shunt capacitance, C_{p1}, largely representative of the casing followed by the series inductance element of the outer end parts to the external contacting points as represented by L_{p1}. This equivalent circuit does to some real extent actually represent piece by piece the actual physical contributions of the typical package structure and therefore can be relatively invariant, and thus useful, over a wide frequency range, in fact, typically as high as X-band for many varactors. Of course at any one frequency an equivalent T or π circuit with no physical connotation can be generated to represent the transformation of the junction to the external terminals; however, such an equivalent circuit is only useful at one frequency, whereas this filter structure can be useful over several octaves. As in many applications several frequencies may be involved, such as parametric amplifiers and harmonic generators, the desirability of having a very broadband package equivalent circuit is self evident.

It is worth emphasizing here that the useful bandwidth of such a representation is inversely proportional to the LC product much as the LC product determines the cutoff frequency of a low pass T or π filter section. Therefore, it behooves the device designer to choose a package structure which minimizes the LC product and therefore maximizes the useful bandwidth of the device. The L/C ratio, however, can vary considerably with package design as this value is related to the impedance level at which the device is to be used which in turn depends upon the application and junction characteristics desired. Finally, it might be noted that this circuit, too, can be simplified somewhat depending upon the impedance state of the variable capacitive element which may make either C_{p2} or L_{p2} negligible for a specific situation.

The General Transformation Representation: The most general approach to the package representation is of course the three-port transformation matrix such as illustrated in Figure 5.3. Although this representation is the most general, it suffers from two principal drawbacks; one, that the matrix evaluation is valid only at one frequency and two, that the matrix elements need not necessarily have any physical relationship to the parasitic elements of the package structure. On the other hand, at the higher microwave and millimeter wave frequencies

Packaging Considerations

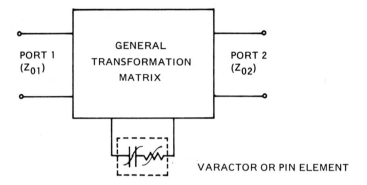

BASIC 3-PORT REPRESENTATION

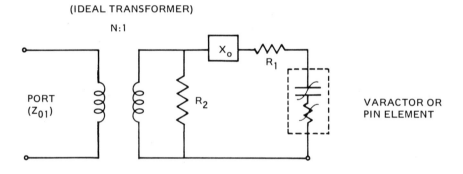

TRANSFORMATION NETWORK
(A SPECIFIC 2-PORT REPRESENTATION)

Fig. 5.3 Microwave Package Transformation

this form of package representation provides the only meaningful form as the package becomes a truly distributed structure of complex geometry and of dimensions significant compared to a wavelength. The input and output ports of this three-port representation are generally considered to be the interface between the package and a uniform transmission line or waveguide such as would exist with a packaged varactor mounted across the center of a waveguide. Under conditions where the packaged structure can be considered small compared to a wavelength in the direction of propagation, it is possible to use the usual approximation in such cases and reduce the three-port structure to a two-port structure with the input and output ports becoming identical. Under these conditions it is sometimes useful to more specifically identify the transformation as is also represented in Figure 5.3. In this specific two-port representation the varactor element at one port is coupled through a resistive and reactive network to an ideal transformer of turns-ratio n to the input-output port at the intersection of the uniform transmission system. For this specific representation, the transformation effects of the package and/or mount on the varactor element capacitance, C, the element resistance, R_s, and the element quality factor, Q_c, can be readily seen for a particular frequency from the following relationships for the transformed varactor and an experimental determination of the transformation network element values $(X_0, R_1, R_2, \text{ and } n^2)$. Assuming $(X_0 - X_c) \gg R_2$, then

$$C_t = C \text{ transformed to external port} = \frac{1}{2\pi f n^2 (X_0 - X_c)} \qquad 5.1$$

$$R_t = R_s \text{ transformed to external port} = n^2 (R_1 + R_s) \qquad 5.2$$

$$Q_t = Q_c \text{ transformed to external port} = \left| \frac{X_0 - X_c}{R_1 + R_s} \right| \qquad 5.3a$$

$$\approx \left| \frac{X_0}{X_c} - 1 \right| Q_c \text{ if } R_1 \ll R_s \qquad 5.3b$$

For any suitably mounted packaged element, it is possible to obtain the general transformation matrix elements or the specific network elements of the approximate two-port representation discussed by appropriate measurements. Evaluation of the transformation network parameters for the two-port representation can be accomplished by utilizing packaged structures with known impedances at the semiconductor or varactor element terminals and making suitable standing-wave measurements in the uniform transmission line with reference to the input-output port. Most frequently-used impedance levels replacing the semiconductor element for reference purposes are open and short circuit conditions at this port together with some known finite impedance such as an appropriate resistor or capacitor or known variations in impedance as obtainable, for example, from the varactor element itself as a function voltage at low frequencies. The various methods appropriate to an evaluation of this two-port representation are described in detail under high frequency measurements contained in the IEEE Standard on Varactor Measurements, Part I — Small-Signal Measurements, as well as other specific references found in the literature.[1]

Packaging Considerations

5.1.3 Package Limitations

In addition to the effects of the parasitic elements of the package in coupling the variable capacitance diode or varactor to the external circuitry, other electrical limitations or restrictions are possible as a result of packaging. These include the problems of arcing and fusing of internal connections. More specifically under certain conditions arcing either internal or external to the package casing can occur; for example, with very close spacing arcing can occur between the strap making contact to one side of the junction and the base portion of the semiconductor element or its mounting pedestal. In addition, it is also possible to have arcing occur across the outside of the ceramic or insulator casing between the metal end pieces, particularly for very thin ceramic spacers. These arcing problems, which may be brought on by high bias or applied RF voltages, generally do not exist for well-designed packages for specific diode units; however, this restriction does exist for any given package structure and can be a problem for high-breakdown voltage units placed in extremely small packages. Indeed, individual varactor units or PIN's possessing breakdown voltages of several hundred volts or special series stacked units contained within a small microwave package are often limited by such arcing problems. As such arcing, particularly as it exists internal to the package, usually brings on the transport of metal so as to worsen surface conditions, this type of package limitation usually results in irreversible changes and device failure and therefore must be avoided.

In addition to arcing, another form of package limitation which also results in permanent damage or failure is that of fusing or opening up of the junction contact strap or wire mesh. In some packages, in an effort to reduce the internal parasitic capacitance, such small cross-sectional area contact straps or wires are used that, when high RF current pulses or forward bias currents are employed, fusing of the strap occurs before damage to the semiconductor element itself. Again, device failure by package limitation rather than by failure of the semiconductor element itself can exist. Thus each package structure possesses both of these limitations and it is important in encapsulating a given unit to match the specific varactor element specifications to these package characteristics as well as to the extent of electrical transformation produced by the parasitic elements.

5.1.4 Special Packaged Structures

Before leaving the subject of the electrical properties of various package structures utilized to encapsulate variable capacitance diodes, it is important to note that there are many special unconventional package structures which are also employed. In an effort to avoid or minimize the transformation effects of surrounding parasitic elements provided in the conventional package, varactor elements are often integrated directly into the circuit structure into which they are to be incorporated forming an integrated structure or package which provides a minimum of transformation between the semiconductor element and the transmission line or waveguide circuitry of the component. In these so-called integrated structures, the semiconductor element is generally mounted in such a way that the metallic connections or supports are made directly part of the uniform transmission line with no ceramic casing involved. In this manner it is possible to reach the minimum possible LC product for the connections and adjacent shunting capacitance such that the upper cutoff-frequency of the mounting configuration can reach well up into the millimeter wave region. To bring the uniform transmission line or guide structure close to the junction so as to minimize the series inductance, it is generally useful to transform the main interaction line or guide impedance down to a very low impedance level with its accompanying close spacing. Sealing of such a unit can be accomplished in two ways, either by suitably pacifying the surface of the semiconductor element prior to enclosure in the integrated circuit structure or by using suitable coaxial or waveguide windows to provide hermetic sealing of that section of the circuit structure. As operation at higher and higher frequencies, particularly in the millimeter and possibly into the sub-millimeter wave region, is pursued, together with the desire to obtain small compact microwave circuit structures, still more emphasis will be placed on the evolution of fully integrated semiconductor diode elements in circuitry also miniaturized by various deposition techniques on high dielectric constant substrates.

5.2 Thermal Properties

5.2.1 Package Thermal Resistance

Of primary importance in determining the power handling capability of a packaged varactor element is its ability to dissipate power. As all dissipated heat must flow out of the dissipation region of the semiconductor element through the package to the external circuit structure and heat sink, the internal temperature rise above that of the heat sink or ambient condition is determined by the thermal resistance of the package and the heat flow through this resistance. It is therefore important to identify the contributions to the thermal resistance of the package and its appropriate thermal model. A detailed discussion of the effects of this package thermal resistance including its evaluation and appropriate model is taken up in the next section on power handling capabilities. However, it is pertinent here to point out, relative to the typical package presented in Figure 5.1, that the thermal resistance of the typical package is largely contributed by the post or pedestal mounting structure within the package and the strap or mesh structure which forms the connection to the opposite end, provided of course that the diode mount adequately incorporates the diode pins into a still

Packaging Considerations

larger thermally conductive path or structure, if indeed not a heat sink. It thus becomes important to minimize these thermal resistance contributing elements of the package by making the lengths of these metal parts small and the cross-sectional areas large. It will be noted, of course, that these are the same directions one must go to minimize the package inductance effects and so these two parasitic elements, one electrical and one thermal, are both minimized by the same geometrical changes. Typical values of the package thermal resistance will range from a few tenths to tens of degrees centigrade per watt.

It should also be noted that in determining the power handling or dissipation capability of these diodes, the package represents only a part of the *total* conductive path from the dissipating region within the semiconductor element to a heat sink external to the package. Thus, the package thermal resistance contribution will be lumped in with the conductive portions of the semiconductor element to yield the total thermal resistance of the packaged diode element. In addition, particularly for transient thermal behavior, this conductive heat path will be assumed to be approximately one-dimensional; this assumption, as can be seen from Figure 5.1, is at least approximately valid for the contribution of the package as it is primarily due to the pedestal or post structure internal to the package.

5.2.2 Package Thermal Time Constant

To completely describe the thermal character of the package, it is required that in addition to the thermal resistance, a fundamental thermal time constant also be determined in conjunction with a suitable model. Again, the detailed model which can be used will be discussed in the following section, but it is of value to point out here that the time constant of the conductive heat path is proportional to the square of the path length (assuming a one-dimensional model) and, of course, dependent upon the material thermal parameters composing the package parts. Thus the shorter the path length, or in this case the post height, is made in order to reduce the thermal resistance, the faster the thermal response will be for this section of path length. Typically the time constant for the total conductive path including that of the semiconductor element is of the order of 1 to tens of milliseconds. In addition, the package holder for most transient considerations can be considered to be an effective heat sink at a temperature determined by the average heat dissipated. Actually the mounting or holder structure as connected to the external ambient will generally have some thermal resistance and associated time constant which, being in the order of seconds, is sufficiently long compared to most transient conditions of interest to be considered a heat sink. Because of the desire to possess a long thermal time constant for the conductive heat path, which itself must be kept short to minimize the temperature drop, the semiconductor element is often top loaded at the junction strap contact by adding a block of material possessing high heat capacity comparable at least in size to the semiconductor element so as to act as a peak heat storage element to minimize the rapid temperature response of the element to pulse dissipation. This so-called top loading has the effect therefore of smoothing out high transient temperature peaks and in effect lengthening the thermal time constant without adding to the thermal resistance of the total diode structure.

REFERENCES

1. W.J. Getsinger, "The Packaged and Mounted Diode as a Microwave Circuit," *IEEE Trans. on Microwave Theory and Techniques MTT-14*, 58-69 (1966).

BIBLIOGRAPHY

Blake, C. and Dominick, F. "Transmission Test Method for High Q-Varactors," *Microwaves, 4*, 18-23 (January 1965).

DeLoach, B.C. Jr., "A New Microwave Technique to Characterize Diodes and 800 Gc Cutoff Frequency Varactor at Zero Volts Bias," *IEEE Microwave Theory and Techniques, MTT-12*, 15-20 (January 1964).

Harrison, R.I., "Parametric Diode Q Measurements," *Microwave Journal, 3*, 43-46 (May 1960).

Houlding, N. "Measurement of Varactor Quality," *Microwave Journal 3*, 40-45 (January 1960).

POWER AND SIGNAL HANDLING CAPABILITIES

6

6.0 General

Having discussed the physical operation, electrical characterization and packaging effects of variable capacitance diodes including the PIN, it remains remains to discuss the signal and power handling limitations of these diodes. In particular, the operating restrictions in using the variable capacitance diode from both the small and large signal points of view are briefly considered together with a presentation of the dissipative or thermal considerations for these diodes. The relationships between these operating limitations and the diode structure and use are stressed.

6.1 Signal Considerations

6.1.1 Small-Signal Limitations

Many applications for the utilization of variable capacitance diodes or varactors as an adjustable circuit element exist, including electronic tuning, filtering, phase or frequency modulation, phase shifting, etc. In each of these applications the varactor is used as a linear element whose value is independent of the signal applied but is adjustable by means of a dc or ac bias voltage. Thus in these applications one is interested in the average value of capacitance existing at a given bias voltage. However, because of the nonlinear relationship which generally exists between the small-signal capacitance and the voltage, small-signal amplitude limitations exist if the biased varactor is to be considered a linear circuit element. Such small-signal limitations are discussed for each of the two bias states.

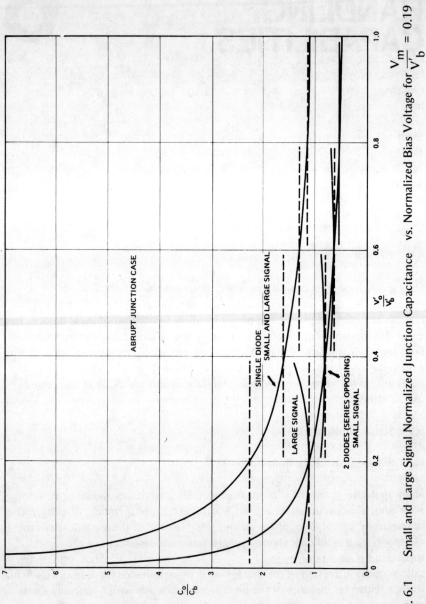

Fig. 6.1 Small and Large Signal Normalized Junction Capacitance vs. Normalized Bias Voltage for $\frac{V_m}{V'_b} = 0.19$

Power and Signal Handling Capabilities

In the reverse bias region, a P-N junction varactor generally possesses a nonlinear relationship between the capacitance of the junction and the applied bias voltage such that as the amplitude of an ac signal applied to this capacitor at a given bias point is increased from an infinitesimal to some finite value, the average value of capacitance seen by the signal source changes (generally increases). Thus for any of the applications mentioned above, the component or circuit becomes signal level sensitive as depicted in Figure 1.1 (applications) and also in Figure 6.1 (single diode curves) where the instantaneous capacitance deviation from the zero signal level bias value is illustrated (extent indicated by the projection of the dotted line on the C-V curve for $V_m/V_b' = 19\%$).

Such a nonlinear characteristic results not only in a change in the average capacitance seen, but of course also introduces nonlinear signal distortion which in effect distorts the signal waveform by creation of harmonics for the single sine wave case. Such harmonics can be analyzed using the capacitance coefficients given for the time-dependent capacitance representation of the diode. Using the capacitive coefficients given in Figure 4.1, it is possible to define the percent change in average capacitance and the percent distortion in signal waveform, both as functions of the relative signal voltage amplitude as compared to the bias voltage. The expressions for these two effects of C-V law nonlinearity are given below.

$$\% \text{ Change in Capacitance} = \Delta \bar{C} \cdot 100 = \frac{\bar{C} - C_0}{C_0} \cdot 100 = (a_0 - 1) \cdot 100 \qquad 6.1$$

$$\% \text{ Cosine Wave Distortion} = \frac{\left(\sum_n C_n^2\right)^{1/2}}{C_0} \cdot 100 = \left(a_1^2 + a_2^2 + \text{--}a_n^2\right)^{1/2} \cdot 100 \qquad 6.2$$

where $a_n = f(\gamma)$, $\gamma = \frac{V_m}{V_0'}$.

Using these two equations and the permitted value of nonlinearity, the restriction placed on the small-signal utilization of the contemplated circuit or component can be determined. Illustrated in Figures 6.2 and 6.3 (single diode curves) are examples of the percent change in signal capacitance and distortion respectively as a function of bias voltage (expressed as a fraction of the breakdown voltage) for a fixed signal amplitude (19% V_b). For this illustrative example, if a variable capacitance diode is utilized in the reverse bias state with a relative voltage swing of 50% ($V_0'/V_b' = 0.38$), the change in average capacitance seen is 6% and the distortion in signal is 30%. For a relative amplitude swing of 25% ($V_0'/V_b' = 0.76$) values of 2% and 14% are found to exist. The importance of the percent distortion can be diminished in some instances where filtering of the harmonics produced are largely eliminated by inherent or intentional filtering action. However, in those instances where the waveform is more complex, such as that of a modulated carrier, cross-modulation products can be produced within the desired bandwidth

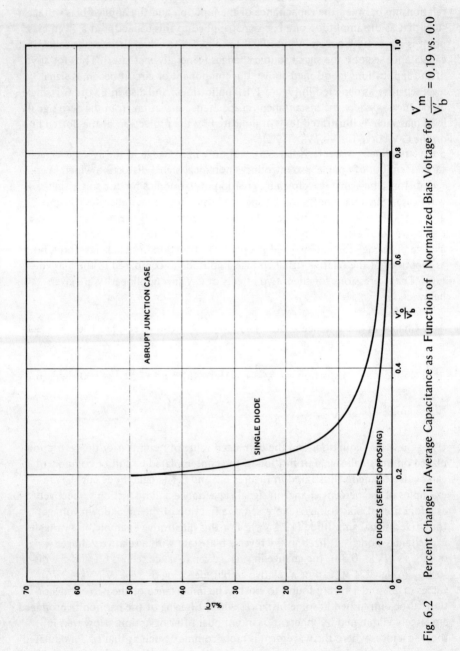

Fig. 6.2 Percent Change in Average Capacitance as a Function of Normalized Bias Voltage for $\frac{V_m}{V_b'} = 0.19$ vs. 0.0

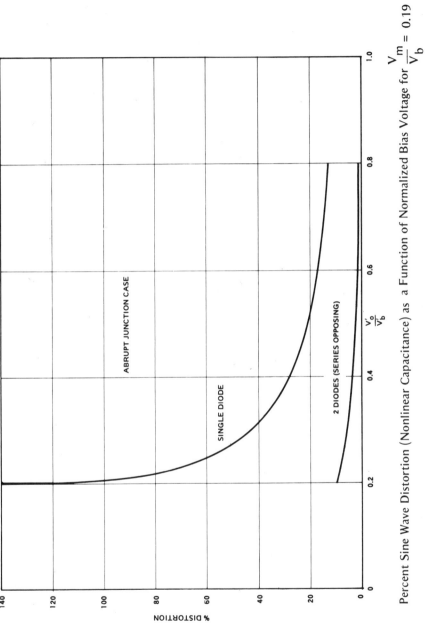

Fig. 6.3 Percent Sine Wave Distortion (Nonlinear Capacitance) as a Function of Normalized Bias Voltage for $\frac{V_m}{V_b} = 0.19$

of the component such that the distortion cannot be diminished or eliminated and becomes a major limitation. It should be noted that these limitations, expressed in terms of capacitance change or waveform distortion, are dependent upon the relative signal amplitudes with respect to the bias applied. Thus on an absolute signal scale it is important to use large bias voltages in conjunction with diodes having large breakdown voltages. However, it must be recognized that the total capacitance change or adjustment obtainable with bias voltage is correspondingly diminished.

Although the average capacitance change and distortion problems discussed above are representative of the general situation encountered, there are two situations in which both of these effects may be diminished or eliminated. First, as was pointed out in Section 3.1.1, it is possible to obtain C-V laws which are nearly linear, at least over a portion of the voltage range, by suitable fabrication of the P-N junction such as the hyper-abrupt junction described. Such variable capacitance diodes, at least over the range of voltages where the capacitance of the junction is directly proportional to the applied bias voltage, will ideally yield no change in average capacitance and a reduced harmonic content (only fundamental present) with signal level provided the amplitude of this signal is restricted to the linear region. In this instance then, the voltage range over which the diode provides such a desirable characteristic can become the small-signal restriction or limitation. A second situation exists in which a significant reduction in capacitance change and distortion can be achieved for a given signal level by combining two nearly identical variable capacitance diodes. By combining the capacitance provided by two junctions oppositely phased with respect to the applied signal and assumed to be biased at the same point, it is possible to obtain a significant reduction in the ac signal nonlinearity of the composite C-V characteristic for the signal voltage instantaneously reduces the total voltage applied to one junction while increasing that applied to the other with a corresponding cancellation (partial) in capacitance changes. This technique can be quite useful in many applications but does have the added complexity of requiring two junctions and suitable bias and signal circuits. The marked improvements obtained in this 2-diode mode of operation are illustrated in Figures 6.1, 6.2 and 6.3.

The variable capacitance diode can also be used as a controllable capacitive circuit element when biased in the forward direction. However, this is generally not done because of the shunt conductivity which also accompanies the diffusion or injection capacitance as cited in Section 3.1.2. This shunt conductance, appearing across the junction capacitance, provides losses in addition to the series ohmic losses such that the total equivalent series resistance is higher in this state which, together with the much larger effective junction capacitance, provides a much lower Q capacitive element as a variable circuit element. In addition, as the diode is biased in the forward conducting state, noise power is produced such that an

Power and Signal Handling Capabilities

equivalent current noise generator appears in shunt with the junction capacitance. This noise is generally appreciably higher than the Johnson noise associated with the ohmic resistance, particularly from VHF frequencies down to the audio range and can be particularly detrimental when employing the element in a low noise circuit. Nevertheless, if a variable capacitance diode is suitably biased with a current source, it is capable of providing large capacitive values (some 2 to 3 orders of magnitude higher per unit area than the barrier capacitance). Furthermore, as was indicated in Section 3.1.2, the diffusion or injection capacitance which normally constitutes nearly all of the junction capacitance at currents corresponding to current density values of 1 amp/cm^2 or more is approximately linearly proportional to the current such that no change in average capacitance and reduced waveform distortion is obtained with signal level as described before for the linear C-V characteristic.

The small-signal limitations on the PIN as a variable capacitance diode are similar to those of the varactor in the reverse biased state except that punch-thru generally occurs at a small fraction of the breakdown voltage as noted in Section 4.3.1. Thus only the voltage range from punch-thru to zero bias is useful for providing a variable capacitance element as bias voltages greater than punch-thru provide an essentially constant capacitance element. A constant capacitance element is also obtained even for voltages less than punch-thru if the operating frequency is above the base or I-region dielectric relaxation frequency (several tens of megahertz) as the device capacitance is then determined by that of the I-region independent of any junction depletion region therein contained.

Little, if any, variable capacitance is gained with the PIN in the forward biased state for the conductivity modulated base or I-region resistance becomes the dominant device impedance with the junctions becoming effective short circuits. In this state, therefore, the PIN is more nearly a variable resistance element shunted by the capacitance of the base region as bounded by the degenerate (metallic) P$^+$ and N$^+$ regions. This variable resistance property, however, can provide a valuable circuit element for attenuator and modulator components.

6.1.2 Large-Signal Limitations

In utilizing the P-N junction as a variable capacitive element or varactor, the operating region must be restricted so as not to encounter high conductive losses, noise generation or diode failure. These limitations, which are usually encountered in large-signal applications, generally rule out or limit the voltage swing into either the forward biased state or into the breakdown region.

With the application of an applied signal voltage to the varactor such that excursions into the forward biased region occur, charge injection will take place; this can be both an advantage and disadvantage depending upon the frequency of operation and the application at hand. Some charge injection can be useful in augmenting significantly the capacitance change developed provided the charge injected is nearly totally recovered as the applied voltage reverses polarity such that little or no rectification or conduction losses exist.(See Section 4.2.) Under such large-signal or hard pumping operating conditions, it is often possible to obtain unusually high reverse currents produced by the impact ionization and resultant multiplication of carriers retrieved with reverse bias following injection. Such high reverse biased currents can lead to an apparent "breakdown" condition well below the actual breakdown voltage of the diode. This condition, which becomes most noticeable with large injection angles (fraction of voltage period exceeding zero bias) such as found in limiter and some harmonic generator applications, can lead to severe dissipation and resultant diode failure. Another limitation on the utilization of such injected charge or injection capacity is the noise generated as charges traverse the junction region. The shot noise so generated puts a severe limitation on the usefulness of injection capacity in low-noise parametric amplifier applications. Finally, it should be noted that heavy injection and conduction is generally highly desirable in limiter applications where the minimum possible large-signal impedance is required to obtain maximum isolation. Under such large-signal, zero bias conditions the conductivity modulation produced in the base region (depending on lifetime) can be significant in reducing the equivalent series resistance which is the dominant impedance element in this state of operation. The extent to which conductivity modulation may be useful in this operation is dependent upon the reverse current buildup with signal level as discussed above.

As noted previously in Section 4.3.1, to operate the PIN in the capacitive state does not require confinement of the RF voltage to the reverse bias region as for the varactor. At frequencies above several hundred megahertz for most PIN's, the capacitance of the diode is essentially invariant with both frequency and RF voltage swing provided zero or reverse bias is applied. With forward bias, the base or I-region conductivity modulation (electron-hole plasma) is established with its resulting resistance state. At microwave frequencies, little change in this state is induced with changes in the RF current level because of the rapid electric field reversal and little net carrier motion. At the lower microwave frequencies (typically, S-band and below), some increase in the value of the forward biased resistance state can be observed with very high RF currents (10^2 to 10^3 times the bias current) but generally thermal failure of the device will occur first, particularly at the higher microwave frequencies.

Power and Signal Handling Capabilities 111

The breakdown region, unlike the forward injection region, provides no advantages in operation. With signal excursions into this region the very least that will result is a sharp increase in the conduction losses coupled with avalanche noise generation. More generally the diode is apt to fail by virtue of the high dissipation encountered as a result of appreciable conduction through a very high field region (resulting from the application of voltages of the order of V_b). Thus the varactor or PIN is seldom, if ever, used with voltage excursions into the breakdown region which thus becomes an operating boundary or limitation.

The above operating restrictions or limitations are illustrated and enumerated in Figure 6.4 for the case of a large-signal sine wave applied to a varactor V-I characteristic.

6.2 Dissipative or Thermal Considerations

6.2.1 Temperature Limitations

Aside from breakdown voltage restrictions, most power limitations are ultimately determined by the temperature rise and the associated physical changes within the diode itself. As the temperature attained within a diode structure is increased to the order of several hundred degrees centigrade, alloying metals and soldered connections begin to melt, some eutectic points between metals and semiconductors are reached, and appreciable diffusion (by the faster diffusants) begins to take place, not to mention problems with organic and some glass encapsulations as well as thermal stresses. Although the temperatures typically required for irreversible changes of these kind to take place are generally high (>250°C), these temperatures in many instances can be reached without having the ambient conditions and/or signal dissipation cause such values to exist directly. Under reverse biased conditions as the diode temperature is increased the leakage current can increase to the point that it contributes significantly to dissipation and further elevates the internal temperature causing thermal runaway to occur. The temperatures necessary for such a thermally unstable condition to result are generally significantly lower than the destructive temperatures themselves, e.g., 75-120°C for germanium, 125-175°C for silicon, and over 200°C for gallium arsenide diodes. Thus, temperature values somewhat below these values are generally listed as the maximum temperatures to be attained for reliable operation. Of course, still lower temperatures may be specified for a given operation to restrict the variation in one or more of the electrical parameters of the variable capacitance or PIN diode.

Since, in general, there is no one specific location in or near the semiconductor element for which a failure temperature can be specified, the *maximum temperature* rise within the diode is utilized in conjunction with a set operating tempera-

112   VARIABLE CAPACITANCE DIODES

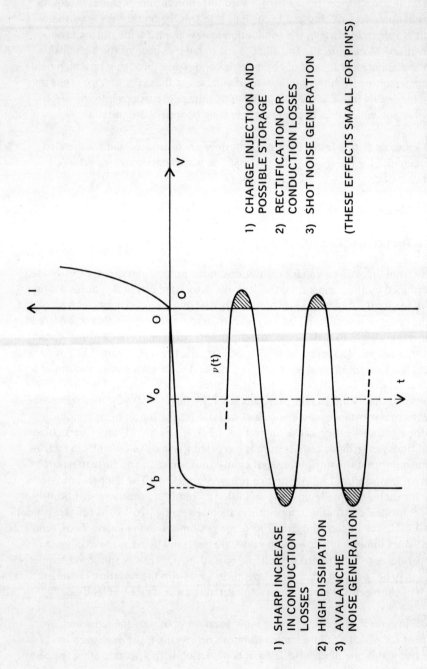

Fig. 6.4 Operating Region Restrictions for Large Signal Use of Varactors or PIN's in Reverse Bias.

Power and Signal Handling Capabilities

ture limit to establish a design power dissipation limit. Thus it becomes essential to be able to calculate or determine the maximum temperature rise within the element in utilizing varactors or PIN's in high power applications. The following sections describe the sources of dissipation, the resulting thermal models, and the calculation of temperature rise.

6.2.2 Sources and Locations of Dissipation

Dissipation or I^2R losses in the varactor or PIN resulting from the flow of dc bias and/or RF or microwave current through the device can be located and evaluated provided knowledge of the effective resistances and rms currents are known. To determine the physical volume (necessary to establish the thermal model and maximum temperature) occupied by the resistances as well as their effective values, it is necessary to first define the various operating conditions from which resistances and resulting dissipation in two principal locations are found.

The most common operation of the variable capacitance diode consists, of course, of bias and signal excursions within the reverse biased state (less than either the breakdown voltage, V_b, or the punch-thru voltage, V_p). Under such conditions the principal resistance (and dissipation) within the device is located in the base region as discussed in Section 3.2.1 and is simply the average ohmic resistance of the region, $\bar{R}_b$(rev.), where averaging is required over the signal cycle if R_b is time (voltage) dependent as noted in Section 4.1.2. Dissipation is likewise located in the base region (or I-region in the case of the PIN) if bias and signal excursions exist in the forward biased state (switching and limiter operations), but the ohmic resistance, $\bar{R}_b$(fwd) or R_π for the usual PIN is generally less in value because of conductivity modulation (resulting from injection) and is determined principally by the bias current (resistance typically not time-dependent by signal voltage with frequencies much above 100 MHz). DC or bias dissipation in these operating conditions is generally negligible in the reverse biased state but can be significant in the forward biased state and should be added to the base region, signal dissipation.

Dissipation can be principally located in one or both of the junction or transition regions (P^+N and N^+N for the varactor, $P^+\pi$ and $N^+\pi$ for the usual PIN) under certain operating conditions. For example, switching operations in the reverse biased state at voltages greater than the punch-thru voltage (assumed to be less than V_b) results in a depleted base or I-region and resistances, R_b or R_π (punch-thru), largely located in the transition regions. Another example is operating conditions with voltages existing in the reverse biased state, but at sufficiently low frequencies such that shunt junction leakage dissipation is most significant. Thus two principal locations, the base (equivalent to the I-layer for PIN's) and junction regions, exist where heat can be generated by dissipation within the semiconductor element. Dissipation or heat generated in the additional contributions (solders, internal metal contacts and mounts, etc.) making up the total equivalent series resistance is generally small and connected with little thermal resistance to the effective heat sink that such dissipation can be neglected in determining the maximum temperature rise within the device.

6.2.3 Thermal Models

Since the heat dissipated in the varactor or PIN is distributed differently in the various operating conditions, a thermal model for each of the conditions would normally be required. However, as discussed above, dissipation occurs primarily in only two locations, in the base and junction or transition regions. Thus only two thermal models need be considered, together with the added fact that dissipation can exist in one or both junction regions, provided that the power dissipation appropriate to the chosen operating condition is utilized. Two such thermal models are now presented.

One useful model for dissipation in the base or I-region has been taken from the physical structure of the device and is considered to consist of a region corresponding to the base or I-region of length, 2W, with a cross-sectional area, A, in which heat is uniformly dissipated during the course of an applied, power step function. (See Figure 6.5a.) This region is considered bounded by the two junctions which are connected by thermally conductive, one-dimensional paths to an appropriate heat sink assumed constant for the time intervals of interest. For the sake of analysis the structure is assumed to be symmetrical about the center of the base or I-region where the maximum temperature, $T_b(t)$, in the unit will exist. From physical considerations this assumption of symmetry seems reasonably valid, at least for the flow of heat for the time intervals typically of interest (transient response), and, in any event, effective thermal parameters should be determined by measurement as well as by calculation which in turn should yield the correct value of the maximum temperature from the results of this analysis.

Two questions arise concerning the validity of this model, namely, the one-dimensional flow and the uniformity of the heat dissipation throughout the base or I-region. Regarding the one-dimensional heat flow, it is physically evident that within the base region (typically a cylinder), and in fact in some instances the semiconductor wafer as a whole, that the flow of heat must indeed be essentially one-dimensional provided uniform heat dissipation exists within the region as there is no means for significant lateral or radial flow. This is not typically the case, however, for planar units (diffused or Schottky barrier type) where lateral flow is possible. Nevertheless, the one-dimensional model can yield approximate thermal results even for these units provided effective, one-dimensional values for

Power and Signal Handling Capabilities 115

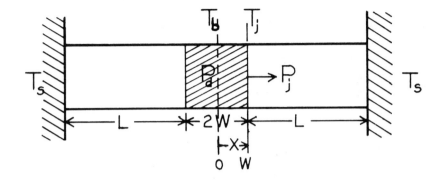

a) Forward Biased State

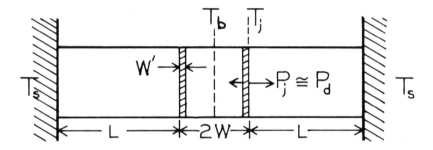

b) Reverse Biased State

Fig. 6.5 Thermal Models for Varactor R-PIN Diodes with Dissipation in (a) the Base or I-Region and (b) the Junction Regions

the thermal resistance and time constant (calculated or measured) are used. The conductive path external to the semiconductor element itself generally changes gradually enough in cross-sectional area to permit it to be represented by a one-dimensional model which, at worst, will yield slightly higher temperature responses than those which will actually exist if there is some degree of lateral heat flow.

The question concerning the uniformity of heat dissipation arises from the fact that there can be a radial dependence of RF or microwave current flowing through the base or I-region of the diode both because of skin effects and the existence of a nonuniform carrier density within the base region. With regard to skin effects within the base region, the resistivity values used for the base region of most varactors range from 0.01 to 1 ohm-cm. Such a range of resistivity permits a conduction skin depth of approximately 3.5 to 350 mils at 1 GHz. As most varactors of appropriate capacitance value to be used at this frequency which employ the base resistivities given would have junction radii smaller than these skin depth values, the current distribution, at least as determined by skin effects in the base region, would be essentially uniform. This conclusion can be generalized to include most operating frequencies provided the diode junction diameter is appropriately scaled. The PIN, with its near-intrinsic base or I-region, will never exhibit in normal use such base skin effects when reverse biased. Even when forward biased, with currents typical of switching operations, resistivities of 1 ohm-cm or lower are achieved and the typical PIN unit will only begin to exhibit skin effects in the I-region at X-band and above.

Another skin effect influence on the distribution of current flow in the base or I-region concerns the distribution of current as it enters the base region from the metallic or degenerate semiconductor regions of the diode. Since these regions definitely possess resistances limited by skin effect above several hundred megahertz, then the currents at these frequencies proceeding to the base region from the metallic parts flow into the semiconductor element from the outside surfaces of these parts and must distribute the current uniformly through the base region even though the currents enter from the outer perimeter or circumference of the semiconductor element. If the impedance associated with the skin effect limited current flow in the metallic or degenerate semiconductor regions becomes appreciable compared to the base region impedance for currents flowing from the outermost part (circumference) inward to the center of the semiconductor element, then the current flow will not be uniform within the base region even if skin effects within the base region itself do not exist. An approximate calculation of such impedances reveals that values of as much as several hundredths of an ohm can exist for typical elements at X-band. Since this impedance is only in the order of 5-10% of the total series resistance values observed for these diode structures, the dissipation can be considered to exist approximately uniformly in the base region. In addition to skin effects, nonuniform carrier densities can exist in the base or I-region because of nonuniform injection, geometry, and high surface recombination velocities. For such diodes, the uniform model will, at best, be roughly representative. In conclusion, considering both the question of the uniformity of dissipation within the base region and the one-dimensionality of the heat flow, it is apparent that the thermal model proposed for dissipation in the base region, although hardly precise, is a reasonable first-order model for most varactor and PIN structures suitably scaled in size for a chosen operating frequency.

Power and Signal Handling Capabilities

The model proposed for dissipation in the junction or transition regions is similar to that for the base dissipation case but differs for the punch-thru state in that the heat dissipated within the base region is not generally uniform but is largely concentrated at the boundaries or junctions.(See Figure 6.5b.) In fact, the principal loss in this state (within the semiconductor element) exhibited by the varactor or PIN must be located in the transition regions between the P^+ and N^+ regions(degenerate and relatively lossless) and the base region (N or π) which is assumed to be totally depleted (and thus free of conductive losses). For dissipation largely due to junction leakage (with no punch-thru condition) where the base dissipation can be neglected, this model is also applicable except that dissipation will occur primarily at only one transition region or junction, the rectifying junction. To the extent that the junction leakage dissipation is uniform, the one-dimensional model is valid. As a substantial portion of the leakage current for many varactors and PIN's can occur at the junction surface or circumference, however, the temperature rise derived from this model will yield only a rough approximation (on the *low* side) for this operating condition.

As the transition regions are generally of the order of 0.1 to 0.3 mils or less, the fundamental time constant for these dissipative regions can be shown to be less than 0.1 μs for silicon units. Thus, if power transient responses or pulse widths greater than 0.1 μs are anticipated, the heat dissipated in the two junction regions can be considered to be dissipated in planes acting as one boundary of a conductive, one-dimensional path to an appropriate heat sink. This model, too, can be considered to be approximately symmetrical for analysis purposes, but, in this case as $T_j(t) \geqslant T_b(t)$, the maximum temperature rise within the unit can be found by determining $T_j(t)$.

6.2.4 Calculation of Temperature Rise[1]

Base Dissipation Case: Based on the model described in the previous section, an analysis of the temperature response has been made by solving the heat flow equation with its appropriate boundary conditions assuming a step input of uniform power dissipation within the base region and the fact that the junction temperature is a relatively slow function of time compared to that of the base. The heat flow problem was divided into two spatial regions, that is, the region of dissipation or base region and the region of heat conduction from the dissipation region to the effective heat sink. The maximum temperature obtained in both of these regions was determined and of course is located for the model chosen at the center of the base region and at either junction which forms the boundary between the two spatial regions mentioned. Since both of these temperatures are important in that the maximum temperature within the base region represents the maximum temperature anywhere in the device and the junction temperature is the temperature which can normally be monitored, both of these temperatures as a function of time are presented.

The temperature response, $T_b(t)$, at the center of the varactor or PIN base region with respect to the junction temperature, $T_j(t)$, for uniform, step power dissipation in that region is as follows:[1]

$$T_b(t) \approx \frac{P_d \theta_w}{2} \left(1 - \frac{32}{\pi^3} \sum_{n=1,3,...} \sin \frac{n\pi}{2} \frac{e^{-n^2 \frac{t}{\tau_w}}}{n^3}\right) + T_j(t) \qquad 6.3$$

or in approximate closed form by determining the value of the summation for small values of t/τ_w.

$$T_b(t) \approx \begin{cases} 0.405 \, P_d \, \theta_w \, \frac{t}{\tau_w} + T_j(t), & 0 \leq \frac{t}{\tau_w} \leq 0.23 \qquad 6.4a \\ \\ 0.5 \, P_d \, \theta_w \, [1 - e^{-t/\tau_w}] + T_j(t), & \frac{t}{\tau_w} \geq 0.23 \qquad 6.4b \end{cases}$$

where the symbols used are as follows:

P_d = power dissipated in the base region in watts

θ_w = thermal resistance of the two halves of the base region effectively in parallel (assuming symmetrical heat flow) in °C/W = W/8.36 $k_w A_w$ (W = half width of base region, k_w = base heat conductivity, and A_w = effective base cross-sectional area).

τ_w = fundamental thermal time constant of the base region (from center) or the region whose length is W in sec. = $4W^2/\pi^2 \alpha_w$ (α_w = base thermal diffusivity)

From Equations 6.3 or 6.4, the temperature difference between the center of the base region and the junction, $T_b(t) - T_j(t)$, is seen to initially increase linearly with time (as would be expected from energy considerations without heat conduction) followed by an exponentially decreasing rise to a *steady-state* value equal to the product of the power dissipated times 1/2 the thermal resistance of the base region (two halves in parallel).

As an illustration of the relative temperature rises obtainable, consider a silicon diode which possesses a base or I-region thickness of approximately 3 mils and a diameter of 14 mils for which the thermal resistance, θ_w, calculates to be 0.3°C/W

Power and Signal Handling Capabilities

and the fundamental time constant, τ_w, to be 7.2 μs. With these thermal parameters and assuming an RF or microwave dissipation of 200 watts, the following temperature rises at the center of the base region are calculated with reference to the junction temperatures. For a time interval of about 100 μs, which is approximately 14 times the time constant, the temperature reaches the maximum difference of 30.0°C; for 10 μs, a difference of 22.5°C is obtained and for a 1 μs interval, a 3.3°C difference is obtained. It is evident from this example that, depending upon the time interval, time constant, and thermal resistance, there can be very significant differences between the temperature monitored at the junctions and that existing within the base region itself. This temperature difference must be determined and added to any measured (junction) values of temperature rise to obtain the maximum temperature within the diode structure.

With the knowledge of the heat flow within the base region due to a step in power dissipation, it is possible to obtain the junction temperature rise due to the outflow of heat through the conductive path bounded on the base region end by the junction and on the far end by an effective heat sink at temperature T_s. The resultant approximate expressions for the junction temperature response in three time regions are as follows:[1]

$$T_j(t) \approx \begin{cases} \dfrac{\pi P_d \theta_\varrho}{8} \dfrac{t}{\sqrt{\tau_w \tau_\varrho}} + T_s, & 0 \leqslant t \leqslant 1.62\,\tau_w & \text{6.5a} \\[6pt] P_d \theta_\varrho \left(\dfrac{t - 0.81 \tau_w}{2 \tau_\varrho}\right)^{1/2} + T_s, & 1.62\,\tau_w \leqslant t \leqslant 2\,\tau_\varrho & \text{6.5b} \\[6pt] P_d \theta_\varrho + T_s, & t \geqslant 2\,\tau_\varrho & \text{6.5c} \end{cases}$$

where the symbols used are as follows:

θ_ϱ = thermal resistance of the conductive path from each of the junctions in parallel to the effective heat sink in °C/W = $L/8.36\,k_\varrho\,A_\varrho$ (L = effective length of the path, k_ϱ = path equivalent heat conductivity, and A_ϱ = effective path cross-sectional area)

τ_ϱ = fundamental thermal time constant of the path whose length is L in sec. = $4L^2/\pi^2 \alpha_\varrho$ (α_ϱ = path equivalent thermal diffusivity)

These approximate relations indicate that the thermal response of the junction initially increases linearly with time followed by a period in which it continues to rise at a less rapid rate corresponding to the square root of time until a time in excess of approximately twice the fundamental thermal time constant of the conductive path is reached whereupon the temperature reaches a *steady-state* value simply given by the product of the power dissipated times the thermal resistance of the conductive path plus the effective temperature of the heat sink. In this regard, it should be noted that the conductive heat path usually consists of a number of sections of different material and cross-sections, so that in practice, if only one thermal resistance and time constant is utilized, it can be expected that there will be a slow change (generally with a time constant of the order of seconds) in the effective heat sink temperature, T_s. This effective heat sink temperature will thus typically depend upon the average power dissipated compared to the instantaneous or peak values and can be significantly higher than the ambient temperature.

To illustrate the junction temperature response for the same circumstances selected before and assuming a fundamental time constant of 7.2 ms and a thermal resistance of 3°C/W for the path between the junction and the effective heat sink, then from Equation 6.5a, it can be calculated that for a 10 μs interval or pulse, a temperature rise of 10.3°C is obtained and for a 1 μs pulse only a tenth of that or 1.0°C is obtained as, for these time intervals, the temperature rises linearly with time. For a pulse width of 100 μs, Equation 6.5b is used to obtain a temperature rise for the junction of 48.5°C for the conditions cited. Thus the maximum temperature within the diode element under these assumed conditions for a pulse width of 100 μs becomes the sum of the temperature difference between the center of the base region and the junction equal to 30°C plus the temperature rise between the junction and the effective heat sink, which is 49°C, plus the temperature of the effective heat sink, T_s, which might typically be 50°C, for a total temperature of 129°C; this approaches the reliable limit of operation for most silicon varactor units. For the particular diode illustrated here, 200 watts peak dissipation power appears to be near the dissipation limit for pulse widths of 100 μs with substantially greater dissipative powers possible for shorter pulse widths allowing approximately the same maximum temperature rise (e.g., for a pulse width of 1 μs, P_d = 4 KW).

As noted earlier, the preceding temperature response expressions are based on the assumption of a relatively slower increase in T_j than T_b with the application of a power step or pulse. Where this condition is not appropriate, a multiplicative correction factor (determined from reiterative solutions), $1/(1 + \beta)$, can be applied to the temperature expressions 6.3, 6.4 and 6.5a [modifying (T_b-T_j) and (T_j-T_s)] where

$$\beta = \frac{\pi^3}{32} \sqrt{\frac{\tau_w}{\tau_\ell}} \frac{\theta_\ell}{\theta_w} = \frac{\pi^3}{32} \frac{k_w}{k_\ell} \sqrt{\frac{\alpha_\ell}{\alpha_w}} \frac{A_w}{A_\ell} \qquad 6.6$$

Thus if $\beta \leqslant 0.1$, all the thermal response relations are approximately correct as derived. An evaluation of Equation 6.6 yields a value of $\beta \approx 0.4\, A_w/A_\ell$ for the one-dimensional model assumed with silicon and copper in the W and L regions

Power and Signal Handling Capabilities

respectively. Thus if the ratio of areas for the base region and conductive path are at least 1/4, $\beta \leqslant 0.1$ and the assumption employed of a slowly varying junction temperature with time (with respect to $T_b(t)$) is valid. Since, in general, the ratio of areas is considerably smaller than 1/4, all of the approximate relations given are generally appropriate. For the situations where β does exceed a value of about 0.1, the correction factor must be applied as noted along with the following time modifications: To Equation 6.4b, a delay factor is included such that the exponent becomes $\frac{t-\Delta}{\tau_w}$ to account for the effective delay in the rise of temperature to its final equilibrium value (caused by the smaller thermal gradient in the base region with the increasing T_j) with Δ being defined as

$$\left[\ln \frac{1}{1.23\,(1+\beta)} + (0.23 + 1.23\beta) \right] \tau_w$$

Also the time boundary between Equations 6.4a and 6.4b is delayed from $t/\tau_w = 0.23$ to $t/\tau_w = (0.23 + 1.23\beta)$. To Equation 6.5b, the delay factor is increased by $(1 + \beta)^2$ becoming $0.81\,(1 + \beta)^2 \tau_w$ and similarly the time boundary between Equations 6.5a and 6.5b becomes $1.62\,(1 + \beta)^2 \tau_w$. An example of the basic thermal responses, $T_b(t)$ and $T_j(t)$, for a step input of dissipation as represented by Equations 6.4 and 6.5 is illustrated in Figure 6.6 for a $\beta = 0.3$, $\theta_\varrho/\theta_w = 10$, and $\tau_\varrho/\tau_w = 1000$.

From the above temperature expressions, it is possible to obtain the waveform of the transient thermal response for any power dissipation waveform by convolving it with the thermal impulse response (derivative of above step responses). For pulse dissipation with sufficient time for complete cooling to T_s between pulses (off time $> 2\tau_\varrho$), the temperature response or waveform can be obtained by taking the difference in the response due to two power step functions, initiated one pulse width, δ, apart. Figure 6.7 illustrates the complete thermal pulse responses for two different pulse widths of base dissipation using the diode thermal parameters cited in the examples earlier. It should be noted that the sharp rise and square-like junction temperature response obtained for narrow dissipation pulses ($\sim 1\,\mu s$) are not generally observed experimentally, but rather a less rapid rise giving the appearance of a continued temperature rise following the dissipation period. However, both the period of the thermal overshoot and the maximum temperature reached appear in reasonable agreement with calculated results.

Junction or Transition Dissipative Case: Using the thermal model described earlier for the dissipation in the transition or junction regions (see Figure 6.5b) and considering pulse dissipation in excess of $0.1\,\mu s$ duration, it is appropriate to evaluate the junction temperature (which in this state is also the maximum temperature) treating the power as dissipated in a plane. Again solving the one-dimensional heat flow equation with the appropriate boundary and initial con-

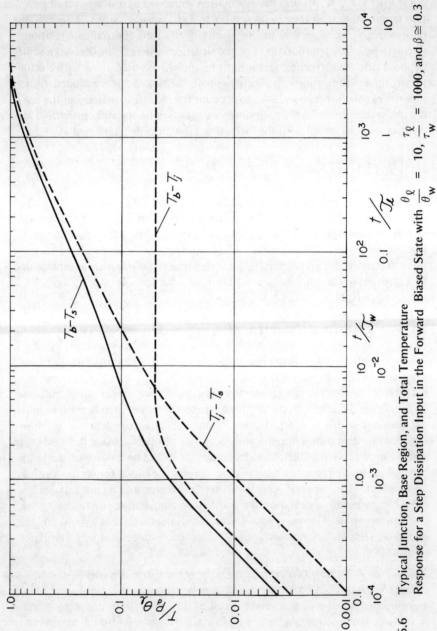

Fig. 6.6 Typical Junction, Base Region, and Total Temperature Response for a Step Dissipation Input in the Forward Biased State with $\frac{\theta_Q}{\theta_W} = 10, \frac{\tau_Q}{\tau_W} = 1000$, and $\beta \cong 0.3$

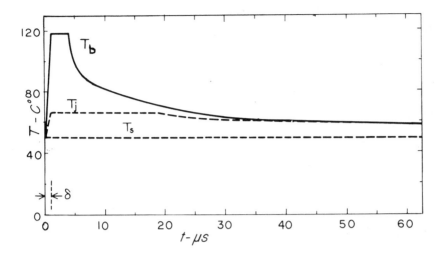

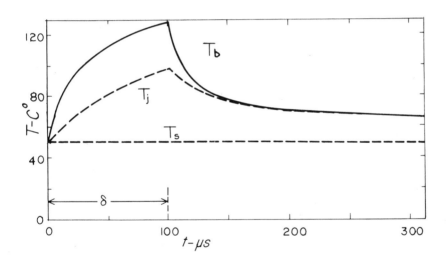

Fig. 6.7 Calculated Temperature Response for Base Region, Pulse Dissipation with Diode Parameters $\theta_w = 0.3°C/W$, $\theta_\varrho = 3°C/W$, $\tau_w = 7.2\,\mu s$, and $\tau_\varrho = 7.2\,ms$ — (a) $P_d = 4\,KW, \delta = 1\,\mu s$; (b) $P_d = 200\,W, \delta = 100\,\mu s$

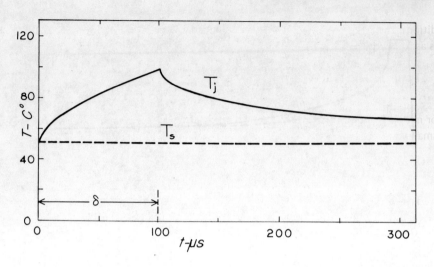

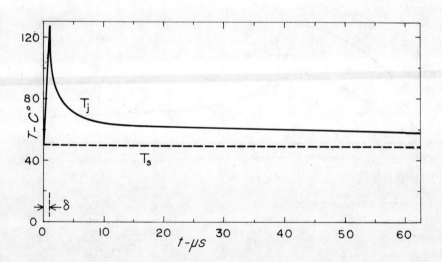

Fig. 6.8 Calculated Temperature Responses for Junction Region, Pulse Dissipation with Diode Paramters $\theta_w = 0.3°C/W$, $\theta_\ell = 3°C/W$, $\tau_w = 7.2\,\mu s$, and $\tau_\ell = 7.2\,ms$ — (a) $P_d = 4\,KW$, $\delta = 1\,\mu s$; (b) $P_d = 200\,w$, $\delta = 100\,\mu s$;

Power and Signal Handling Capabilities

ditions, the junction temperature step response for the junction dissipation case is as follows:[1]

$$T_j(t) = \gamma(t) P_d \theta_\varrho \left[1 - \frac{8}{\pi^2} \sum_{n=1,3...} e^{-n^2 \frac{t}{\tau_\varrho}} \right] + T_s \qquad 6.7$$

or in approximate closed form by determining the value of the summation for small values of t/τ_ϱ,

$$T_j(t) \approx \begin{cases} \gamma(t) P_d \theta_\varrho \left(\dfrac{t}{2\tau_\varrho} \right)^{1/2} + T_s, & 0 \leq t/\tau_\varrho \leq 2 \qquad 6.8a \\ \\ P_d \theta + T_s, & t/\tau_\varrho \geq 2 \qquad 6.8b \end{cases}$$

where all parameters were previously defined except for $\gamma(t)$ which is a time dependent factor less than or equal to unity accounting for the small heat flow (for storage) into the base region adjacent to the junction when dissipation occurs at both junctions or transition regions. With dissipation principally at only one junction $\gamma \equiv 1$ and θ_ϱ and τ_ϱ must be adjusted to reflect the addition of the base width, W, to the effective, conductive heat path, L. An approximate expression for γ for two junction dissipation case is given below:

$$\gamma(t) \approx \frac{1}{1 + \beta\sqrt{\dfrac{3\tau_w}{t+3\tau_w}}} \qquad 6.9$$

which is smallest when $t = 0$, $(1/1 + \beta)$, and increases to unity as $t \gg 10\,\tau_w$. Again, it is pointed out that generally $\beta \leq 0.1$ such that $\gamma \approx 1$, but the complete expressions for temperature, Equations 6.7 or 6.8, with 6.9 provide approximate evaluation even for $\beta > 0.1$. The pulse responses for the reverse biased dissipating cases are illustrated in Figure 6.8.

Comparing the junction temperature responses for forward and reverse bias cases, Equations 6.5 and 6.8, it is seen that the response for the reverse bias case is much faster initially (up to $t \ll \tau_w$) because of the concentration of dissipation in the junction region. Thus, except for situations for which θ_w is excessively high (extremely wide base region or smaller cross-sectional area), the *maximum* temperature within the device will be higher in the junction dissipation case for short pulses (in this case $< 2\,\mu s$) with the same dissipation, P_d, as the added temperature rise (T_b-T_j) will be insufficient when added to Equation 6.5. However, for longer pulses, the added temperature drop in the base region will add significantly such that the maximum temperature reached in the device will always be higher for the base dissipation case assuming equal values of dissipation.

Based on the thermal models and relationships presented here, it should be possible to ascertain the maximum temperature reached within a diode element for any of the various operating conditions and dissipation waveforms. The thermal parameters required for such an evaluation of T_{max} can be approximately calculated for a given, mounted diode (assuming its detailed structure is known) or measured using the thermally dependent electrical characteristics of the junction together with the functional, temperature vs. time relationships given here. With the knowledge of what maximum temperature is tolerable and the predicted thermal response for a given diode element and operating conditions, it is possible to specify both the peak and average thermal power limitations.

REFERENCES

1. K.E. Mortenson, "Analysis of Temperature Rise in p-i-n Diodes Caused by Microwave Pulse Dissipation," *IEEE Trans. on Electron Devices, ED-13*, 305-314 (May 1966).

BIBLIOGRAPHY

Hines, M.E., "Fundamental Limitations in RF Switching and Phase Shifting using Semiconductor Diodes," *Proc. IEEE, 52*, 697-708 (June 1964).

Moroney, W.J. and Brunton B.H., "Temperature Rise of PIN Diodes from X-band Pulse," *Electronics Letters 1*, 165 (1965).

INDEX

Arcing, 99
Biasing, 10, 13
    forward, 18-21, 23, 77, 109-110
    reverse, 27, 32-33, 62, 105
Breakdown voltage, 25, 32, 83
Capacitance
    depletion layer, 10-15, 27-40, 62, 77-78
    diffusion, 21, 33, 52
    injection, 18, 21, 23-25, 33, 52, 78-79, 110
Circuit configurations, 3-6
    pumped mode, 4-5, 110
    parametric amplifiers, 5, 61
    tuning, 3-4, 103
Conductivity modulation, 25-26
Contact potential difference, 9
Cut-off frequency, 46-47, 48-50, 69
Dielectric relaxation, 16-17, 44
Diodes
    snap, 81
    abrupt junction
        definition, 1-2
        doping profile, 10
        depletion layer, 13
        capacitance of, 28-40
        breakdown voltage, 32
        figure of merit, 69-71
    graded junction
        definition, 2
        doping profile, 10
        depletion layer, 13, 15
        capacitance of, 28-40
        breakdown voltage, 32
        conductance, 56-59
        figure of merit, 69-71

      hybrid, 2
      byper abrupt, 2, 30-32, 108
      PIN, 3, 24, 81-89, 109, 110-111
Doping profile, 1, 10, 16, 28, 83
Effective operational resistance, 17-18
Equivalent circuits, 94-98
    circuit models, 79-81, 84-86
      current determined charge evaluation model, 61, 77-79
      time dependent capacitance evaluation model, 61, 62-64, 75-76
Figure of merit, 69-75
Frequency dependence, 15, 61, 81, 88-89
    capacitive, 16-17, 21, 23, 33-34
    resistive, 24, 41, 44, 67-68
Fusing, 99
Gain, 61, 64, 84
Heat dissipation, 113-114
    base, 113, 117-118
    junction, 117-118, 121, 125
    thermal models, 114-117, 118, 121
Integrated structures, 99-100
Losses, 15-16, 17-18, 25
Noise, 24, 69-70, 108-109
Parasitic elements, 40, 91-94
Q values or quality factor, 3, 18, 21, 46-48, 58-59, 69, 108
    cut-off frequency of, 48
    determination of, 25
    frequency of, 98
Recitification, 23
Resistance losses, 2, 15-16, 17-18, 21, 25-26, 40, 75-76
Signal distortion, 105, 110
    elimination of, 108, 110
Temperature limits, 111, 113, 125-126
Temperature response, 15, 34-36, 44-45, 76
Time response, 15, 33-34, 101-102
Transfiguration matrix
    2-port, 98
    3-port, 96-98
Varactor measure, 47